To my grandchildren.

LANGUAGE:

FROM SENTIENCE TO COGNITION

RAFAEL PINTOS-LÓPEZ

CONTENTS

Foreword	vii
1. Introduction *A Cartesian, meta-evolutionary understanding of human consciousness*	1
2. The Language Roadmap *Towards a solution*	22
3. A Historical Perspective *Judaeo-Christian references to consciousness*	105
4. A Contemporary Perspective *Adam Frank, Marcelo Gleiser & Evan Thomson*	142
5. Another Philosophy *Eastern schools of thought*	148
6. Japan and Zen	167
7. Materialism vs Dualism *Robert Lawrence Kuhn*	175
8. Conclusion	178
Acknowledgments	187

RAFAEL PINTOS-LÓPEZ
Copyright © 2025 by Rafael PINTOS-LÓPEZ
All rights reserved.
No part of this book may be reproduced in any form or by any electronic or mechanical means, including information storage and retrieval systems, without written permission from the author, except for the use of brief quotations in a book review.

 Formatted with Vellum

FOREWORD

It was a late September day in Minato-ku in central Tokyo. The street bore the stillness and sweet scent which followed the third typhoon of the season. It was early morning, but clearing and cleaning was in progress. The broken tiles over the entrance to the Buddhist temple were being repaired, priests at the neighbouring Shinto shrine were putting order back into their garden. I crossed the canal into Higashi-Azabu, and towards Shiba Koen park. Stopping at a drink machine, I chose a hot coffee. The bottle warm in my hands, the coffee, pure pleasure. I turned at the sharp sound of a can being kicked along the pavement. An elementary school boy: short sleeves, yellow sun hat, over-sized black leather backpack, was contemplating his prey. However, his opportunity proved short-lived, as he was scolded by an elderly passer-by. Picking up the can, he deposited it in a bin. He smiled at me. I smiled too. We shared meaning.

∼

This book delves into questions of consciousness. What is it like 'to be'? How is it that we are conscious of being? How is it possible we can share meaning between two minds? Is this the result of a 'natural' physical evolutionary process, or something different? Do other sentient creatures feel the 'unbearable lightness of being'. Do other sentient creatures consider themselves 'special'? Is our superior view of ourselves merely down to human ego?

Humanity has pondered these questions, since it could ponder. Many celebrated minds receive a mention in this book, including: Plato; Aristotle; St Augustine; Democritus, Copernicus, Descartes; Wallace, Darwin, Wittgenstein; Schopenhauer; Heisenberg, Schrödinger; and Chomsky. Nonetheless, despite this trove of combined intellect to build upon, modern neuroscientists and philosophers continue to find a definitive theory of human consciousness elusive.

The current dominant paradigm is one that follows Darwinian logic. Materialism, or 'physicalism' if you like, makes sense to our modern brains. We are, after all, of our time. There is comfort in knowing that science has, or will, eventually come up with the answer – it is a matter of discovering the right equation. Nevertheless, it is difficult to escape a nagging feeling, from the inside, that we are more than a collection of molecules, of neurons. Surely, we *H. sapiens*, are special, we have 'something higher'? Surely, we possess an element of the transcendental?

This book questions the orthodoxy on consciousness. It suggests we are looking for answers using the wrong philosophical approach. The text contends that language and culture have a larger impact on individual and collective consciousness than they are generally credited with. More specifically, it is proposed that human consciousness comprises two distinct elements: sentience, essentially the result of physical evolution, and high-level cognition, the product of evolution of language and culture – 'cognitive evolution', if you will. This dualistic approach to consciousness, a separation of body and mind, is not a new idea, as students of Descartes will attest, but the arguments around the relative contribution language and culture are. The book delves into many potential ramifications of such a hypothesis.

Notwithstanding this hypothesis, the real importance of this book may not be limited to the theories it contains. Certainly, they appear to have an evolutionary logic, a particular plausibility, which could provide an alternative explanation for human consciousness. Of equal value, however, may be the suggestion that we focus attention to possibilities other than materialism – that we suspend our Darwinian conditioning and use a wider, interdisciplinary, lens to explore a fresh perspective on the 400-year-old theory of dualism proposed by the father of modern philosophy.

If this book does not fit the mould, neither does its author, my friend, Rafael Pintos-López. He is, by his own admission, an oddity. An Argentinian-Australian, Rafael was born in Santiago del Estero, in 1942. While a product of post-war Argentina, there has always been more than a hint of the Age of Enlightenment about him. Argentina avoided the destruction in Europe and the Pacific wrought by the Second World War, and perhaps this allowed Rafael to maintain a stronger connection to a classical historical tradition. Whatever the reason, Rafael has the ability to think differently. He is professionally a linguist, creatively a painter, logically an inventor, and spiritually a romantic. He defies categorisation. Fortuitously, his breadth of intellect also makes him well placed to ponder the interdisciplinary question of consciousness.

Rafael's inquisitiveness is infectious and the book is full of both interesting questions and interesting propositions. I'd invite you to read it with an open mind.

J Watts

INTRODUCTION

A CARTESIAN, META-EVOLUTIONARY UNDERSTANDING OF HUMAN CONSCIOUSNESS

"Here's to the crazy ones. The misfits. The rebels. The troublemakers. The round pegs in the square holes. The ones who see things differently. They're not fond of rules. And they have no respect for the status quo. You can quote them, disagree with them, glorify or vilify them. About the only thing you can't do is ignore them. Because they change things. They push the human race forward. And while some may see them as the crazy ones, we see genius. Because the people who are crazy enough to think they can change the world are the ones who do."
Apple Advertisement-
Steve Jobs

Monk: *"What is the difference between the enlightened and the unenlightened man?"*

Master: *"The unenlightened see a difference, whereas the enlightened do not see such difference"*

Neuroscientists and philosophers are currently finding that a theory of [human][1] consciousness is an extremely elusive concept. There is no universal agreement on a definition of consciousness. The subject presents a series of complications that most researchers find baffling.

The truth—and this will be explained in detail—lies hidden in plain sight. Some of the everyday activities *H. sapiens* individuals do, which are considered 'human nature', are not actually natural or biological at all. They are intertwined with truly biological phenomena, but they are, in reality, artificial skills that need to be taught.

Some of the questions that puzzle philosophers nowadays are: How can something non-physical, like experience, emerge from something purely physical, like the brain? Can we explain the electrical correlates of thought? This book maintains that those questions are part of the wrong philosophical approach.

What scientists and philosophers are trying to find out is the connection between our subjective perception of reality and the individual human brain. I propose that if we want to find out about human consciousness, we have to follow a different path altogether. In that sense, this book is an attempt to challenge convention.

But there is much more. We live a human reality that has grown in unexpected ways. The problem is we have not realised it as yet. Humanity functions as a superorganism. How can that be? We would have noticed, surely. Well, it has been happening for so long, that it seems natural now. But superorganisms are not purely biological units. Termites use

---

1. The square brackets point towards a taxonomy that will be explained below.

basic communication and instinct. The tool human individuals use to function as a superorganism, to co-operate, is complex communication. We use language and a [hybrid][2] consciousness. We can do things termites cannot do. Termites can build mounds. We are a polymathic species. We can do many things.

Is thought possible without language? Yes, it is. We can see that many other animal species solve basic problems. They have a basic type of thought, a germ of thought. But we should make the distinction. There are qualitative differences. Human thought is definitely more complex than animal thought. Hominins had basic thought before they acquired language. After language, thought grew exponentially. Like never before. How did that happen?

Conducting research on other animal species to find out about human consciousness does not appear to be taking us anywhere. What is evident is that our species has changed to an incredible extent while all other species have remained almost the same. The questions we have to ask ourselves are: Is building a skyscraper possible without language? Is the Internet possible without language? Is diplomacy possible without language? Is philosophy possible without language? Is measuring possible without numbers? Is algebra possible? The answer to all those questions is a resounding no.

Some will claim causality: yes, but all those thoughts are generated by neurones in human brains; consciousness is physical, consciousness is material. The answer is no. Why is humanity different from all other animal species? We have gone beyond evolution. That is ridiculous, they will say.

Did people not ridicule Erasmus Darwin when he suggested

---

2. The square brackets point towards a taxonomy that will be explained below.

we had evolved from something much more primitive? Did society know then that creation was a myth? The good doctor had noticed that horses and donkeys had similarities. He discussed 'transmutation' of species. Dr Darwin said that maybe all life had evolved from a common ancestor. The eighteenth century was not ready for him, or for an answer.

The last few decades have been a period of stasis and self-censorship. There appears to be no questioning of the philosophy behind the search for consciousness. The trend in the research community leans heavily towards materialism. Without issuing a judgment on materialism as a philosophical choice, it is possible to state that—in terms of consciousness—it has led science and philosophy nowhere thus far.

Famously, Professor Chalmers asked the question *"... why does all this feel like something from the inside?"*, and coined the phrase *"The hard problem of consciousness"* back in 1995[3]. Again, in 1998 he wagered neuroscientist Christof Koch that the mystery would not be solved by 2023. He won the bet. The taboo firmly established by the current scientific orthodoxy against anything outside of materialism has resulted in a dearth of original ideas. There has been no breakthrough. Three decades of 'same old, same old'.

Two main problems are involved here: 1) believing that consciousness is a solely biological phenomenon which emerges from the individual brain; 2) rejecting *'in limine'* any explanation that does not fall within materialism. But, above all, science should be inquisitive. Science should be open to

---

3. Chalmers, D (1995) - *Facing Up to the Problem of Consciousness, Journal of Consciousness Studies*, 2(3):200-19.

any type of inquisitiveness. Science should accept original ideas above bias. This has not been happening lately.

These days there are hundreds of theories. Human consciousness—their theories tell us—is part of a general blurry 'consciousness', a seemingly universal concept. None of the theories take into account the history of *H. sapiens*—of our species—as something of particular significance. Many of them appear to assume that consciousness is a fundamental element of the universe, or that, at one point in history, consciousness appeared out of nowhere. They are all based on a subjective, static perspective.

Most theories confuse 'consciousness' with 'sentience', and are based on the individual brain—human or non-human—as a physical entity; they choose to ignore the intersubjective connectivity of the human mind. Lab rats do not communicate like human beings. How can we explain human consciousness experimenting with rats? Many studies attempt to compare human consciousness—an utterly complex phenomenon—with the basic 'consciousness' of other animal species. Some clearly claim that consciousness is a fundamental element of the universe, such as space, or mass, or charge. In desperation, they end up conjuring panpsychism and other, non-falsifiable, solutions. Well, the scientific community should know not to trust anything unfalsifiable. Karl Popper must be turning in his grave.

What is intended here is to demonstrate that the exponential growth of human culture, society and technology has been only possible because of the uniqueness of a human consciousness that has surpassed the rules of biological evolution.

Whether the situation the researchers of consciousness find themselves in is the result of political trends like 'wokeism', or the anthropocentric phobia that resulted from Copernicus' work, is completely irrelevant. What is undeniable is that the research community, and humanity in general, merit a complete rethink of human consciousness. Convention needs to be challenged.

This book attempts an explanation of consciousness taking into account the Western perspective of an 'objective reality', but respects—and integrates—some of the holistic views of Eastern philosophy. This book assumes that sentience is a seemingly ineffable, fundamental, phenomenon that occurs in most living creatures. More about that in other chapters.

Our five senses—vision, audition, olfaction, gustation and taction—give *H. sapiens*, and other species, a direct connection to reality. As with other species, our perception of reality may be far from perfect, but it is exactly what we require to function in nature. I also suggest that the issue of sentience could be eventually addressed [through particle entanglement][4].

In discussing sentience, it is possible to mean only the five senses as the basic points of contact between the individual and nature. The five senses refer to the functions of distinct organs like eyes, ears, nose, mouth and fingers. These are all subjective/passive phenomena. The individual receives sensations from his/her direct experience of the world. Even when that reception may include agency of some sort, e.g., the individual may turn his/her gaze towards something specific, or move arms, hands or fingers in order to touch something specific, perception is still passive (stimulus/response). All of these, of course, are processed through the

---

4. The text in the square brackets will be explained below, with more on the possibilities of quantum mechanics.

inner brain. We will see how sentience is possibly the intake component of a quantum process.

Contact with the outside world may include other means. The individual may touch objects with his/her toes; it is also possible to feel heat and warmth, or cold, through the epidermis, the body surface; and these factors may turn to extremes, with physical results, like frostbite or burns. These are all individual, but peripersonal, phenomena, which exclude interoception, i.e., self-awareness, and/or sensations, like hunger, inner pain, inner pleasure, etc.

There are some areas in which sentience and cognition overlap, like feelings of love, hatred, greed, etc. Other animal species also experience these feelings, no doubt; some of them maybe to a lesser, more basic, degree. The possibility of love from a parent bear towards its pup, for instance, is something quite evident. Other feelings, like hatred or greed, may not exist other than in humans and may be more related to cognition. It appears that the overlapping of sensations/ feelings varies in degree according to the case.

When we discuss overlaps, it is also important to mention identity and memory, for example. Other gregarious species, like whales, dolphins and chimpanzees, may have some semblance of identity, which is something probably required to function even in small groups. There is no denying that the human brain reflects a progression from those of other animal species.

There is a component of consciousness, then, that we have in common with other creatures, i.e., sentience. This book submits that human consciousness differs from those of other species, not just qualitatively and quantitatively, but in terms of its nature as well. The basic hypothesis submits that human consciousness is a unique, 'black swan', meta-evolutionary phenomenon that separates us from the rest of life on Earth. How so?

Arguably, *H. sapiens* has a distinct type of consciousness that involves two layers: sentience and complex cognition. This is a taxonomic approach that, I maintain, is essential in order to understand that human consciousness has historically developed into a basically hybrid phenomenon.

While this book focuses mainly on the metaphysical workings of cognition, it also points out that the hybrid nature of human consciousness is clearly reflected on the physical, layered, development of the human brain. I will attempt to analyse some of those physical aspects in further detail, within the constraints of the book.

A basic analysis of the human brain will lead us to understand the human brain as an adaptive organ, but it would also establish that its growth has followed a lengthy layering process. The famous MacLean 'triune' brain points to a reptilian complex—also known as the basal ganglia—, a paleo-mammalian complex—aka the limbic system—, and a neo-mammalian complex, i.e. the neocortex.

For the purposes of the book, a more appropriate taxonomy of those components or layers will result in a basic subdivision of the brain into a biological internal layer (basal ganglia + limbic system) and an artificial, language-induced, secondary layer.

If we follow the historical process chronologically, it would be possible to establish that the original biological layer was already a highly complex quantum system which allowed mammals to perceive, that is, to experience their natural surroundings.

INTRODUCTION

A philosophical analysis of that perception would establish that it occurs as a process, whereby an external object is replicated in the brain as a 'percept', or mental concept.

The external object exists but, without perception, the process is incomplete, i.e., both stimulus and response are required. That means that if a tree famously falls in the forest without a listener, the actual sound is incomplete, i.e., there is an emission, but without reception the sound itself does not occur. The same happens with all the processes related to sentience. Where there is no perception, there is uncertainty. That implies a certain dynamic interconnectedness.

This is where quantum physics can explain something. According to Heisenberg's Uncertainty Principle, both the position and the momentum of a particle cannot be precisely determined. The principle even provides a formula for the amount of uncertainty ($\Delta x \, \Delta p \geq h/4\pi$) where x is position, p is momentum, and h is Planck's constant. Sentience would involve a perception process that would determine the position of a particle the moment observation occurs, and that would also include an entangled particle in the brain. That analysis would clearly determine that the mammalian brain operates as a quantum system, a notion acceptable by both quantum mechanics and Eastern philosophy. Is that a plausible explanation as to how the brain operates? I can only suggest that it is.

What I also propose is that, after a specific inflection point—such as the introduction of communication (language) — another semi-spheroid layer was added to the physical mammalian brain and now covers it as part of the human brain. That layer could be imagined as two-way mirror/magnifying glass: it still allows the mammalian core of the brain to perceive external objects or actions, i.e., to receive the same impulses and produce the same responses through the specific organs required, but it also communicates with other individuals with minds, while appearing as a mirror to those other individuals (theory of mind).

The magnifying glass function of the layer is for the individual: he/she can analyse his/her cognitive processes and enhance them exponentially through the use of unlimited cultural/linguistic resources. Through communication and co-operation, the human mind as been slowly transformed from an individual organ into a cultural one. That could be defined as both, the onset of meta-evolution and of humanity itself. Maybe it is possible to understand now how humanity operates as a superorganism. Human individuals function within a culture but also need the culture in order to remain human. That occurs inter-generationally through language. More on that below.

It is well-known that all cognitive functions are processed in the neocortical areas, especially in the prefrontal cortex but—after the meta-evolutionary inflection point that resulted from the creation of language—other areas of the inner human brain have also developed further. The limbic system, i.e., where emotions and basic memory are processed, has undergone lesser changes.

There is a clear distinction between the areas of the brain where cognitive functions are processed, as opposed to non-cognitive areas. All areas are interconnected and the human brain has adapted itself to work holistically; however—and maybe arguably—the periphery of the brain has undergone distinct and substantial growth after the introduction of language. In that respect, the superior temporal, the inferior frontal, and the supra-marginal gyri are identifiable as linguistic areas. The superior temporal gyrus is where Wernicke's area is located. That is the centre of language comprehension. The inferior frontal gyrus hosts Broca's area, where speech production is processed. Other linguistic functions, like name-processing and phonology may be found in the supra-marginal gyrus.

Henceforth, and in order to gain further clarity, this book will use the terms 'sentience' and 'cognition' where it refers to 'subjective experience' and 'psyche', respectively. The words used could be 'perception' and perhaps 'thought', or some other terms. There is no semantic implication here; the terms 'sentience' and 'cognition' have been chosen arbitrarily and for clarity's sake.

The two components of human consciousness—sentience and cognition—have been inextricably enmeshed for tens of thousands of years, but are, nevertheless, discrete, discernible, entities. The question that comes to mind is: can we clearly establish that they are discrete and that they have different natures? The answer is: yes. That is one of the aims of this book.

These two layers, or components, have different natures: one of them, sentience is, no doubt, biological. For the second one, cognition—the layer that I submit has been added meta-evolutionarily—it would be possible to use the term 'artificial'.

Language is uniquely human and artificial, i.e., hominins developed language in a process that lasted tens of thousands of years. In doing so, they became human. It was a process that involved creation but also mutual change. Language developed artificially, while the humanity of our ancestors grew. One of the conclusions derived from that fact is that there are no 'natural' languages. No other animal species has anything remotely similar to human language. Going back to our topic, the complexity of human cognition can be considered a byproduct of recursive language. Cognition, thus, could be said to be artificial as well. More about that in other chapters.

Where I submit that human consciousness is a meta-evolutionary phenomenon, I refer to its artificial layer, cognition, i.e., human consciousness can also be explained in terms of a hybrid phenomenon. That the acquisition of human cognition was not a normal biological process appears evident from the behaviour of human cultures, societies and institutions. They do not abide by normal, biological evolutionary, norms. It is something that Russel Wallace appears to have understood, and that even Charles Darwin admitted:

"A great stride in the development of the intellect will have followed, as soon as the half-art and half-instinct* of language came into use; for the continued use of language will have reacted on the brain and produced an inherited effect; and this again will have reacted on the improvement of language. As Mr. Chauncey Wright has well remarked, the largeness of the brain in man relatively to his body, compared with the lower animals, may be attributed in chief part to the early use of some simple form of language,- that wonderful engine which affixes signs to all sorts of objects and qualities, and excites trains of thought which would never arise from the mere impression of the senses, or if they did arise could not be followed out. The higher intellectual powers of man, such as those of ratiocination, abstraction, self-consciousness, &c., probably follow from the continued improvement and exercise of the other mental faculties..."[5].

---

5. Darwin, C, (1870) - *On the Limits of Natural Selection*, North American Review, Issue Oct 1870, p.295. * My underlining.

INTRODUCTION

∼

By now the reader probably surmises that what I propose here is grounded on Cartesian—substance dualistic—philosophical principles. That is correct. All of the theories of consciousness based on idealism or materialism (that purport to include human consciousness) have apparent shortcomings. Explaining the theories or their shortcomings does not fall within the aim of this book.

Let us try to summarise how this book proposes that cognition entered the picture as a meta-evolutionary addition to the—until then—purely biological human sentience. There are some other advanced mammalian species that have developed basic communication systems. Good examples of them would be chimpanzees, whales, dolphins, and some corvids. Maybe their communication systems could not be called languages as such, in that they are used in small groups of individuals and only serve limited purposes. Apart from limited, they are mostly linear. Human language is not linear.

There are other species of less advanced individuals that operate as social structures with large quantities of individuals; bees and termites would be good examples of the sort of species we are discussing. As explained, it is possible to think of them as superorganisms.

Human languages began as the ones used by the former type of species: small groups of individuals that required communication for limited purposes. When those groups, or clans, combined, or grew larger through need, or trust, or territorial expansion, oral communication became more and more complex. Then, recursive language made its appearance.

Using a limited set of words—or units—human languages grew in order to express infinite combinations of those words.

Human languages use compositionality, that is, they can combine words and phrases; they can express complex propositions using the meanings of terms and changing the sequence of those terms. The overall meaning of a sentence varies according to the order of those individual meanings. No other animal communication system can do that.

The fact that human language is productive and flexible makes all the difference. Oftentimes, even when unknown items are included, their meaning can be induced from the context of the sentence.

In one of the above paragraphs, I say that animal communication systems are linear; by that I mean that they lack what human languages have: they use layers. Using clauses and subclauses, our languages add infinite depth and complexity to human thought and expression.

Once human communities became tribes—with perhaps hundreds or thousands of individuals—languages grew accordingly. Division of labour and civilisation did the rest. Language and culture became more important than the individual. From then on, evolution would occupy itself with the culture rather than the individual. Cultures would grow. Evolution would cease being biological and become meta-evolution.

I did not build the house I live in. You probably did not either. This may sound trite, but it can be applied to many instances of our human reality. Clothes, health, transport, education. Human beings need the collective. Human beings operate and behave within the collective.

INTRODUCTION

When evolution became meta-evolution, human cultures had become like communities of bees or termites, but with the added advantage of hierarchies, advanced communication systems, and flexible rules and regulations that would grow according to the needs of the collectives. Further expansion was achieved through wars and innovation.

Human culture had left the individual behind. With the addition of cognition—i.e., sophisticated thought and communication—, human consciousness would become a part-metaphysical phenomenon. From then on, to be able to function as part of the collective, human individuals would need a hybrid consciousness.

∽

Philosopher Dan Dennett used to compare Australian termite mounds to Gaudí's Sagrada Familia Cathedral:

"Termites build these amazing castles, these structures that are both beautiful and extremely efficient. Some of them look very much like the wonderful church that Gaudí built in Barcelona, la Sagrada Familia. Here are two structures, a termite castle in Australia, and a cathedral, or church, in Barcelona. They look very similar. They're both artefacts but they're made by completely different kinds of processes. In the case of the termites there's no boss, there's no architect, there's no blueprint, there's no pre-planning. This is all simply the product of basically mindless, clueless little termites following very simple, local rules up this wonderful structure cuts. Gaudí, very much in contrast, had it all figured out in advance, had a blueprint, had manifestoes, had a declarations of what the principles were, and he lorded it over his subordinates, who lorded it over their subordinates, who lorded it over the people that were putting the stones in place, and all the rest, so we have the difference between a very top down, hierarchical, preplanned, intelligent design in the case of Gaudí, and a marvel of unintelligent design in the case of the

termites; now, the termite castle has the structure it has for good reasons, but the termites don't need to know those. That's carpentry without comprehension. Gaudí, at the other extreme, he's got a clear articulation of the reasons why he wants to do it this way rather than that way, and that's a profoundly different phenomenon in nature, and now, the big puzzle is: how do you make a Gaudí type mind out of a brain which is, eerily, like a termite colony? * *Gaudí's brain is 87 billion clueless neurones. Not one of them even knows anything about churches or sunlight or anything and yet you put those 86 billion neurones in a team together, somehow, and you get a Gaudí mind as the result.*"[6]

At first, Dennett's perspective appears like a valid one, but then, as you analyse it, it does not make a lot of sense. From a distance, Australian termite mounds and the Sagrada Familia have some similarities in terms of shape and contour. Then, Dennett, an avowed materialist, compares the termite colony to the cells in Gaudí's brain. What?

The comparison may sound witty, but it is short-sighted on many accounts. When dealing with human consciousness, materialism is also short-sighted. Materialism fails to comprehend that humanity has long surpassed biological evolution. Materialism cannot account for cognition other than within the individual brain and cannot explain sentience (Wittgenstein would say: because it cannot be explained). Sentience is understood through ostension (showing). Humanity operates as a huge, incredibly successful, superorganism. Not a solely biological superorganism, but a metaphysical one as well. Predominantly, a metaphysical superorganism.

It is difficult to see ourselves as minute components of a superorganism. We consider ourselves as autonomous individuals. Of course we are not like neurones, or cells, or

---

6. Dennet, D (2020) – *Sagrada Familia and Termites, The Search Documentary*, transcript of monologue in video clip. * My underlining.

termites. We are autonomous agents to a certain extent. But living without the collective is almost impossible for us. It is difficult to accept that humanity has become a superorganism, but that is what it is, and that is what we are. A superorganism that has taken over the planet.

Human beings are individuals of the same species that interact synergetically in large groups called cultures. Cultures have generally learnt to operate within the whole of the species. Each culture has its own language or languages, country or countries, division of labour, institutions, etc.

As *H. sapiens* individuals, we are supposed to think alike. And, to a large extent, we do. Human cognition is the common medium through which cultures and languages communicate. We even share institutions like the United Nations, the International Court of Justice, the Food and Agriculture Organisation, the European Union, Mercosur, or Brics; and countries use financial and banking agreements to trade in all types of commodities. The exponential growth of human society does not necessarily have any teleological implications. It just means that human beings have co-operatively accomplished incredible feats that individuals could not have achieved by themselves.

Gaudí's brain was not the result of sixty-odd billion neurones. It was the result of Western civilisation. He was of French descent, born in Catalonia of Catalonian parents, and had learnt Catalan from them. Like all Catalan speakers, he could also understand Spanish, and like most Spaniards, he was a devout Catholic. A catholic architect: it comes as no surprise that he built churches.

Gaudí was one architect; but there were many architects when he was alive, and there still are many architects, and like them, there are many other professionals. He only commenced the building of the Sagrada Familia, which continued for decades after his death and is still unfinished. He built many other buildings. You just have to go to Spain, especially to Barcelona, and look around to see them. Termites build mounds. That is all they do.

Also, there are many other cathedrals, like Santiago de Compostela, Chartres, or Notre Dame. We do not know who designed or built many of them. They belong to humanity, like the Pyramids, near Cairo, or the Wailing Wall, in Jerusalem. And there are many beautiful buildings, such as the Guggenheim Museum in Bilbao—by Gehry—or Ronchamp Chapel—by Le Corbusier. Saying these are just artefacts is a disservice to humanity.

The reason for the prevailing materialistic approach—and also the problem it implies—is that human consciousness is, as a whole, deceptively individual and, at the same time, it appears biological. Truth is, one of the components is biological; but, I repeat, the other one, cognition, is an artificial addendum.

Cognition appears individual to the human individual but, within the collective, it operates as a medium. Once we have understood this, it is possible to begin considering how human consciousness is shared by humanity. And by nothing else. Human beings can comprehend and communicate. As a premise, saying that consciousness is universal is easily falsified. You would need to prove to me that a stone not only understands, but that it also communicates.

Perhaps, before discussing the contents of the chapters, it would be interesting to articulate one of the main premises of the philosophy behind this book.

I propose that any valid, falsifiable, theory of human consciousness should comply with the *sine qua non* acknowledgment of: a) the two components of human consciousness; and b) the discreteness of those components due to their different natures. The chapter *The language roadmap* includes a detailed explanation for such a requirement.

An idea would be to submit an equation such as:

$$HC = \alpha + \psi$$

Where HC is human consciousness; α – $αισθητικότης$ *(aisthikótis)* is sentience; and ψ – $ψυχή$ *(psyche)* is cognition. Human consciousness is not equivalent to [$\alpha\psi$]. The two elements cannot be confused with one another. Although there are some areas where overlap occurs, the two components are certainly intertwined, but are not part of a fuzzy reality, of a blurred whole. There is a clear division between those two components. As already stated, they have different natures. They are discrete.

The advantage of this taxonomic approach is that, if we consider HC as a super set, and α and ψ as sets, it is possible to discern between their subsets and clearly determine that the behaviour of those subsets is going to differ from subsets of the other set. For example: α has clear subsets that correspond to the senses, e.g., vision could be α^1, audition α^2 olfaction α^3, and so on; more importantly, it would be possible to determine the origin of any cognitive subset and clearly establish it as a human creation, hence, $\psi > \psi^1$ (where ψ^1 = Identity), say; $\psi > \psi^2$ (where ψ^2 = Memory); $\psi > \psi^3$ (where ψ^3 = Time) ; $\psi > \psi^4$ (where ψ^4 = Measurement); $\psi > \psi^5$ (where ψ^5 = Numeracy), and so on. The development of cognitive subsets can be culturally, chronologically, spatially and physiologically traced.

Again, if we consider that human cognition is exponentially complex and distinct, no other animal species has equivalent cognitive subsets, and some of them may not even be present in some current human cultures. We can certainly state that, while other animals may perceive change, no other animal species requires time-keeping (no ψ^3; ψ^4; or ψ^5).

In terms of human cultures, Hopi culture does not appear to have developed ψ^3 until recently, and Pirahã speakers do not have ψ^4 or ψ^5 either. Some Pacific islander cultures may have limited notions of ψ^5. One of the aims of the book is to establish that human cultures and languages (and there are thousands of them) have developed at different rates. Many grammars tend to confirm this.

The primary aim of this book is to demonstrate the validity of a theory based on substance dualism and on the discreteness of sentience and cognition as necessary components of human consciousness. To that end, I will draw upon the works of historical and contemporary philosophers and scientists, as well as ancient texts and cultures, and retrieve information from a variety of other sources such as TED talks and TV interviews.

Apart from this *Introduction*, the book will be structured as follows:

The Language Roadmap will describe the basic conjectures of the hypothesis; it will discuss sentience and animal 'consciousness' and whether there is an evolutionary continuum; it will delve into the history of language and meaning; it will outline how language and culture are transmitted as opposed to genetic or biological inheritance; it will discuss the introduction of long-term episodic memory, identity, writing and time; it will discuss the concepts of ethics, morals, free will, and happiness vs pleasure within the context of a consciousness in which time plays an important role; will address adventurousness as a characteristic of human endeavour; it will discuss how creativity and arts such as the visual arts, literature, drama and poetry are reflections of human culture.

A Historical Perspective will address the growth of cognition and include Judaeo-Christian references to consciousness in Scripture; it will address the differences between Western views on objective reality and ancient Eastern schools of thought, such as the Vedantas and Zen Buddhism; it will discuss the views of Schopenhauer in that respect; it will analyse the ideas of Ludwig Wittgenstein in his early period, in particular his *Tractatus Logico-Philosophicus*, and of Erwin Schrödinger, as a renowned physicist analysing biology and consciousness in the series of Lectures *What is Life?*.

A Contemporary Perspective will discuss *The Blind Spot – Why Science Cannot Ignore Human Experience*, a recent volume by Adam Frank, Marcelo Gleiser and Evan Thompson.

Materialism will delve into why materialism is not fit for purpose in terms of consciousness.

The *Conclusion* will attempt to tie up all the topics discussed in previous chapters, including different views on human consciousness.

THE LANGUAGE ROADMAP

TOWARDS A SOLUTION

"Thinking outside the brain means skilfully engaging entities external to our heads — the feelings and movements of our bodies, the physical spaces in which we learn and work, and the minds of the other people around us — drawing them into our own mental processes. By reaching beyond the brain to recruit these "extra-neural" resources, we are able to focus more intently, comprehend more deeply, and create more imaginatively — to entertain ideas that would be literally unthinkable by the brain alone."
The extended mind: the power of thinking outside the brain-
Anne Murphy Paul

When I say this idea is just a roadmap, I do it because it is just a logical, holistic, view of what Chalmers called *"the hard problem of consciousness"*. It points the way in a new direction. It does not purport to be a full theory of consciousness. I believe the actual value of it is that it is a view from a different angle, from outside the box, if you like.

I am sure much more competent researchers and professionals with deeper understanding of the individual issues covered by the conjectures being submitted here will be able to develop them (and amend them) to a much higher degree. The issue is very complicated and it will require intellects with a high level of expertise in order to deal with the different points the problem presents; it will definitely involve long periods of multidisciplinary co-operation.

The most important concepts I propose here are: the hybridness of a layered human consciousness, the issue of humanity operating as a superorganism and, possibly, the quantum nature of sentience within the original mammalian brain, plus the artificial addition of the neocortex, onto what became the human brain—which is also quantum. These phenomena, all based on language, can only be analysed—I believe—from a Cartesian perspective.

There is also the paradox of the creation of an individual identity as a result of enhanced communication and co-operation among individuals, and the extraordinary development of human cultures and civilisations brought about by human consciousness.

Many of the questions that the book attempts to answer have been around for many generations. They involve issues such as time, imagination, and the certainty of human finitude, and long-term episodic memory, but the most important one is the one that addresses human subjective experience. The answers have never been quite satisfactory.

Religion attributed the origin of human life and consciousness to creation by a deity or deities, i.e., no further explanation possible. Belief, certitude and dogma were introduced as definitive answers to human inquisitiveness.

Western philosophy has been trying to delve into human consciousness but always encounters the same barrier as science: how can we explain subjective experience? Where is it? Thus far, science and philosophy have not found a solution as to how it operates.

During the past few years—with modern methods, such as new imaging technology—there have been big advances in neuroscience. Why is it that Western science and philosophy still find themselves confronted with an enigma of such proportions?

Western philosophy and religion were based, from their inception, on a division of the human being as 'the observer' and the rest of nature as 'objective reality'. In the case of consciousness, *inter alia*, the problem has to do with that perspective: according to Western philosophy there is an observer and an objective reality.

We shall see that thinkers like philosopher Arthur Schopenhauer, and physicists Erwin Schrödinger and Karl Heisenberg, among others, could appreciate that there were other angles to the problem. Quantum physics regard perception as the end of a process, without which the process is not complete, i.e., it does not exist. For Eastern philosophy a more dynamic, holistic, view is involved. The book will explore some of those angles. But let us go back to our Western philosophical perspective.

Human consciousness has been considered a paradoxical phenomenon by many at least since the times of King Hezekiah of Judea, Plato, St Augustine of Hippo, Democritus, and many other thinkers, more recent in history. Human beings have senses and, at the same time, have developed complex thought processes. Sentience and cognition have coexisted within us ever since we became human. They are intertwined; they work in tandem, and yet they appear incompatible. How does that happen? They complement each other but cognition fails to explain sentience, and sentience is not interested in explanations. Many philosophers and scientists have thought about it. Philosopher René Descartes, among the most lucid of men, asserted: *"I think, therefore I am"*. That established a clear causation. Yes, in order to ask yourself something, you have to exist. That does not answer the fundamental questions of life and consciousness. His substance dualism, however, points towards a solution.

There is life—which in our case involves sentience, or subjective experience— and then there is thought, which we do not consider a soul in the religious sense any more. The concept originated in what the Greeks knew as 'psyche' *(ψυχή)* and we now understand as cognition. One is physical and the other one, metaphysical. Current philosophical and scientific research have blurred the boundaries between sentience and cognition; furthermore, they have confused the issue even more by often treating human consciousness as just sentience, whereas human consciousness involves the two layers we are discussing and those layers are inextricably enmeshed. However, they are also discrete and should be treated as such. Without cognition we cannot begin to discuss human consciousness or ask questions about it. The phenomenon is unique.

Having failed to understand Ludwig Wittgenstein's *Tractatus Logico-Philosophicus*, originally published more than a century ago, in 1921, Western scientists and philosophers are still trying to explain experience. More on Wittgenstein's early work in the following chapters.

Wittgenstein's conclusion is that science—and/or philosophy—may not be able to completely explain human consciousness because they are not equipped to do it, i.e., being cognitive and therefore language-based, they can explain objective reality but not sentience. Sentience is subjective and cannot be analysed objectively. As a biological process, the workings of sentience are probably fully explainable. That will not show the phenomenon of sentience. The capabilities of human language fall short of explaining sentience. The biological, subjective, component of human consciousness is ineffable. There are glaring differences between 'explanation' and 'ostension'. Some things can be explained. Other things need to be shown.

Materialism, as a philosophical school, falls short. If fails to fully understand or explain consciousness. Again, more on that below.

Mary's Video Paradox

In 1982, philosopher Frank Jackson published the now famous article *"Epiphenomenal Qualia"*[1], where he argued against materialism (physicalism) on the grounds that the qualitative feel of experience cannot be explained. He proposed an experiment in which Mary, a brilliant neuroscientist, lives in a black and white laboratory. She knows everything that needs to be known about colour, including all the physiology of vision, light waves, behaviour, etc. All the information on colour is available to her. She does not need to know anything else. One day, however, she really experiences colour and learns something new. That experiment (and another of Jackson's articles) became known as 'the knowledge argument'. The reasoning behind the story and the conclusion are correct. Unfortunately, Jackson failed to

---

1. Jackson, F (1982) - *Epiphenomenal Qualia, The Philosophical Quarterly*, Vol.32, pp.127-136

explain in clear terms the different natures of sentience and cognition. The concepts were still blurry. The result is that many researchers still question the validity of the argument. I believe the allegory is correct: there are ineffable qualia in sentience. Also, the phrase 'the knowledge argument' is a misnomer. The allegory should have been known as 'the explaining vs showing argument'. Sentience can only be shown.

Decades later, a TEDEd video produced by Eleanor Nelsen involves the same story under the title *"Mary's Room: a philosophical thought experiment"*. The video shows how Mary lives until her computer malfunctions and shows the colour red. A voice in off explains: *"Mary is an expert in colour vision and knows everything ever discovered about its physics and biology. She knows how different wavelengths of light stimulate three types of cone cells in the retina and she knows how electrical signals travel down the optic nerve into the brain. They create patterns of neural activity that correspond to the millions of colours"* [2]. The allegory is much more convincing. The paradox of the video is that while showing that sentience can only be shown and not explained, it needs to also convey in words what cannot be just shown to be understood.

Science and philosophy are cognitive. They were created to elucidate. They are equipped to deal with phenomena from a meta-evolutionary perspective. But they cannot explain sensory experience. It's simple.

I consider human cognition as a meta-evolutionary phenomenon because everything that happened once humans

---

2. Nelsen, E (2017) - *"Mary's Room: a philosophical thought experiment"*, retrieved from *TEDEd videos, YouTube,* publicly available online.

became cognitive is different from what happened before then and from what happens to other species. With all due respect, our societies do not work according to a Darwinian model or abide by Darwinian principles. There are institutions and rules that are exclusively human. As humans, we develop identities, a sense of what is good or evil, we create, we keep time, we produce art, we are adventurous, we have free will, ethics and morals. None of that appears in any other species.

Is it possible to have some details as to why neither science nor philosophy can explain sentience? There are answers and possibilities.

Conjectures

These are the conjectures I submit:

1. Human consciousness is a hybrid phenomenon that consists of two integrated but discrete layers or components: an original, subjective, layer (sentience) —which is natural and biological, and an exclusively human layer (cognition)—which is a subsequent addendum, that grew through language; this latter layer is human-made and, as such, artificial. They have two different natures;
2. Physically, the human brain replicates those layers: the mammalian brain is a [probably quantum] system in which sentience is the original component, i.e., sentience is the necessary biological agent needed to complete any perceptual process; the neocortex, also an interconnected [probably quantum} system, is the cognitive addendum that resulted from language;
3. The structure and operation of the individual human brain, however, does not fully explain human consciousness, which is a meta-evolutionary phenomenon that involves communication and

shared meaning, as well as individual sentience and cognition;
4. Because of the dynamics involved in 2) and 3), the only way to cognitively understand the correlates of human consciousness (from a Western philosophical perspective) needs to be based on substance dualism; [quantum physics and] Eastern philosophy may view the issue from a holistic angle;
5. The discreteness of the components of human consciousness is a result of their different natures;
6. Language is artificial and cultural; language and cognition developed through a lengthy process of mutual feedback; human cognition should be considered a byproduct of language;
7. Complex meaning did not exist before language nor does it exist other than in human language and cognition; complex meaning is also a byproduct of language;
8. In any theory of human consciousness, the notion that sentience, by itself, constitutes human consciousness, or that the boundaries between sentience and cognition are blurred only creates confusion; deeming the components discrete and defining the boundaries between them is of the utmost importance for a proper understanding of consciousness, furthermore, it is a necessity;
9. Sentience should be considered as an ineffable, fundamental, phenomenon that occurs subjectively [at the quantum level] in most living creatures; sentience provides certainty to [quantum] phenomena involved in perception; being biological, it does not ask for explanations nor does it give them, hence its ineffability, as human language cannot describe the richness of experience;

10. Language and cognition are artificial, human abilities that need to be taught, to every individual, in a process repeated every generation, as they are transmitted culturally, not genetically. The individual cannot learn them without intersubjective help.

Maybe in the future, a better understanding of quantum physics' particle entanglement phenomena would give us a clearer picture of the behaviour of the senses, which would not constitute a solution to the *'hard problem'* of subjective experience. It would, however, link the wavefunction to the issue of perception as a process. Viewed from a dynamic perspective of interconnectedness, without a perceiver, the perceptual process is incomplete, or inexistent. This would have vast philosophical repercussions.

Currently, Western philosophy and science appear to operate by analysing reality from a rather static perspective. Eastern views of reality—much like those of quantum physics—understand the world from a holistic point of view, as a dynamic, intersubjective processing system, where the sentient being is relatively integrated into the system and is necessary for events to occur.

Epiphenomenalism—the purely physical explanation of consciousness—is an explanation at the classical level, but mental events are not caused by physical events at the classical level. The human brain [which, I repeat, appears to be a quantum system] operates as an organ that requires a degree of interconnectedness in order to produce both, internal and external effects. The human brain operates with 'objective' input (afference) by interpreting that input as the final phase of a perceptive process. Without that relative interpretation, the event does not occur. The moment we accept that the mind (the seat of consciousness) needs interconnectedness to exist, we accept that a solely physical analysis of consciousness cannot be correct, or complete. Neither the mind is the brain, nor does it exist solely within the skull (or the skin). That is only the explanation Western philosophy provided. But then, consciousness is not the perception of experience. It does not occur outside of experience; it is part of the experience itself.

∽

Schrödinger (1944) asserts that reality does not exist twice, that perception is only a relative phenomenon:

"No single man can make a distinction between the realm of his perceptions and the realm of things that cause it since, however detailed the knowledge he may have acquired about the whole story, <u>the story is occurring only once not twice</u>. The duplication is an allegory, suggested mainly by communication with other human beings and even with animals; which shows that their perceptions in the same situation seem to be very similar to his own apart from insignificant differences in the point of view – in the literal meaning of 'point of projection'. But even supposing that this compels us to consider an objectively existing world the cause of our perceptions, as most people do, how on earth shall we decide that a common feature of all our experience is due to the constitution of our mind rather than a quality shared by all those objectively existing things? Admittedly, our sense perceptions constitute our sole knowledge about things. This objective world remains a hypothesis, however natural."[3]

What Schrödinger states clearly rejects the notion that there is

---

3. Schrödinger, E (1944) – *What is Life?* - Cambridge University Press, p.145.

an 'objective reality' as such. What we perceive and the object we are perceiving form part of one reality and one reality only. The interconnectedness of consciousness is a phenomenon that cannot be easily denied.

The dualistic perspective of the process concurs with quantum mechanics in terms of the event only 'happening', or being complete, when the trajectory of the particle is determined by the sensory participant.

The fact that sentience is efficacious, i.e., that it causes effects on physical events, provides evidence against classic materialism as a way of interpreting sentience.

Now, to the second part of the problem. Sentience is something we share with other animal species. Human consciousness as a whole, however, is something much more complex.

No evolutionary, i.e., biological, study of consciousness can depart from the sponge and reach human consciousness. They are not part of the same continuum; the cognitive component adds a different nature to our minds, a qualitative and quantitative leap: human consciousness cannot be understood from an exclusively evolutionary perspective which is, by necessity, individual and static.

Tens of thousands of years ago, there was a point at which evolution gave way to another phenomenon. Fully developed syntax appeared as something exclusively human, which, I submit, is also artificial and meta-evolutionary. Other human phenomena—byproducts—followed language. Through a process that lasted tens of thousands of years, basically, *H. sapiens* created language and language 'created' *H. sapiens*. By then, our minds had become hybrid, part biological and part human creation.

Cognition is only acquired linguistically, through parental and cultural upbringing. It is individually transmitted. This needs to happen every generation, which makes us a highly altricial species. Cognition has always been and still is introduced by means of language and its nature is as artificial as those of culture and language, otherwise we would be born equipped with full linguistic comprehension and speech. As we shall see in more detail below, this is probably the clearest demonstration of the artificiality of cognition. We are born sentient, but not cognitive. No other animal species, not even other gregarious species, undergo such a complex linguistic, cultural and cognitive socialisation process.

Language learning and cognition are complex mental abilities. They differ qualitatively from bipedalism, for instance, which is an exclusively physical skill with some neural implications. Also, bipedalism has already resulted in hip and muscular changes, and in the atrophy of the fingers of our lower limbs, which are now toes. We have undergone a major change because of it. Bipedalism has become second nature. Children learn to walk on two feet by trial and error, much like birds learn to fly. Sometimes, there may be some teaching involved and some instances where children learn to walk on all fours, but these are clear exceptions.

In between, in terms of complexity, there are similar activities that require teaching, some are basic, like bicycle-riding; others have more complex prerequisites, like touch-typing or piano-playing. But here we have to make another distinction as well, because bipedalism, language and cognition are necessary to be a fully functional human being, whereas other, more recent skills, are not requirements. Not all human beings can ride pushbikes or play musical instruments.

Voluntary imagination, creativity, long-term memory, free will, adventurousness, etc., (the list is by no means exhaustive) appear to be exclusively human traits acquired through language and culture, i.e., they do not appear to have existed before complex cognition. They are not evolutionary. Since thought is produced by language and culture, it is obviously influenced by them (as per Whorfian Linguistic Relativity).

The logical conclusion is that there should be cortical and other specialised centres in the brain—<u>newer than any centre that deals with strictly biological phenomena</u>—where cultural developments are processed (e.g., Broca's and Wernicke's). The suffix 'neo' in the term 'neocortex' points exactly to that. It represents a new development in the human brain.

It could be argued that—after the appearance of language and the development of the neocortex—the human brain has created this dual system in which the periphery of the organ acts like mirror towards the outside, and a magnifying glass towards the individual, adding information, and enhancing any basic cognitive processes that may occur in the limbic system, in the inner brain. These purely biological processes, like non-episodic memory, basic ideas and motivations are therefore conveyed by means of neural networks of synapses to the neocortex, where they are further processed and integrated, and end up becoming cortical representations: long-term memories, complex thoughts and strategic voluntary imagination.

Some biologists and neuroscientists appear to guess, to some extent, that there are intertwined components in our consciousness, that there is a duality in our mind, and that that duality is reflected on the structure of the brain. They can even guess a chronological difference between the components. However, unable to conjure up the birth of language as a milestone, they find the phenomenon very hard to explain and reach conclusions that are far from correct. Biologist

Telmo Pievani (2025), from the University of Padua, seems to be one of them:

"Put simply, the first system is old in evolutionary terms and governs quick, automatic responses — whether in routine or emergency situations — and is primarily connected to the amygdala, cerebellum, and basal ganglia. The second system, primarily connected to the prefrontal cortex, is a relatively recent evolutionary development. It governs our most deliberate actions, those that result from careful and slow evaluation of contextual information. We could call it the logical reasoning system, as it handles the careful analysis of concepts, generalizations, principles, and abstractions."[4]

The way I see it, all of these statements of Pievani's are correct. There is an old system and a secondary, newer system, and their chronological appearance coincides with how peripheral their locations are in the human brain. However—according to the author—they are systems that coexist in the brain without rhyme or reason, and there are no discrete boundaries between them. Pievani then— as sometimes happens with researchers—reaches the wrong conclusions:

"Neither system is necessarily more rational or emotional than the other. Both have played a fundamental role in our evolution: first, by providing us with instantaneous evaluations based on experience, which is preferable when the decision has to be taken immediately or is based on a large number of different variables; and second, by offering us the wonders of science and any choice based on reasoned arguments, especially when we are faced with a new problem. We should not consider one to be irrational and the other rational, because reacting instinctively in certain situations is often the most

---

4. Pievanti, T – *Flat Earthers on a Cruise*, The MIT Press Reader, virtual article adapted from the book *Imperfection – A Natural History*, by the same author, 12.08.2025

rational choice. At the same time, however, both can cause us to make enormous mistakes because our actions are frequently generated from an improvised middle ground between the two systems."[5]

It so happens that there is a difference, and an explanation as well; that one of them is more rational than the other, and that the newer component has gone beyond normal Darwinian, evolutionary, development. All of that follows some logic and evidence of it is generally available to anyone interested in human consciousness.

A recent article in *Neuroscience* by Carla Avolio (2025), of the Max Planck Institute, *"Schrews Shrink and Regrow Brains, Offering Clues for Human Diseases"* offers an example of how comparisons that tend to ignore differences between the human brain and those of other animal species result in misguided conclusions:

"Scientists used non-invasive MRI to study shrews that seasonally shrink and regrow their brains, uncovering water loss as the key driver of this rare phenomenon. Despite losing about nine percent of their brain volume in winter, shrew brain cells remain alive, with aquaporin-4 proteins playing a key role in regulating water movement.

This discovery parallels mechanisms seen in human brain diseases, where brain volume loss typically leads to irreversible damage. Researchers hope studying how shrews regrow their brains could inspire novel approaches to treating neurodegenerative disorders."[6]

According to the author of the article, the scientists who are

---

5. Pievanti, T – *Flat Earthers on a Cruise*, The MIT Press Reader, virtual article adapted from the book *Imperfection – A Natural History*, by the same author, 12.08.2025
6. Avolio, C (2025) - *"Schrews Shrink and Regrow Brains, Offering Clues for Human Diseases"*, retrieved from *Neuroscience*, 1 September 2025.

conducting the study hope to apply their findings to the study of human neurodegenerative diseases:

"For the neurologists, the story of what shrews can offer human medicine has just begun. Many brain diseases—Multiple Sclerosis, Parkinson's disease, Amyotrophic Lateral Sclerosis (ALS), and Alzheimer's disease—involve brain volume decline due to water loss. But for humans, this loss progresses in only one direction. ... The next step for the team is to study the second phase of Dehnel's —the brain's regrowth from winter to summer. By doing so, they hope to unlock clues for treating brain diseases."[7]

The comparison is ill-conceived and cannot lead to a reversal of cell loss in Alzheimer's Disease patients. The phenomenon in shrews is the result of the animal's adaptation to the rigours of the cold season. Water loss is a phenomenon that is spread throughout the brain. In humans, it is normally caused by an acceleration of the normal cognitive decline due to the ageing process. Furthermore, shrews appear to continue conducting their lives with relative normality, whereas AD's patients lose cognitive skills through their disease. The loss of cells in AD is mostly related to neocortical functions, such as long-term episodic memory, time, identity, and language in general, and none of them have parallels in shrews.

Time is a human construct that exists only within human cognition, through unlimited voluntary imagination (expectation) and long-term memory (which includes subjective chronological identity and collective perception). The counterpart is that—like in other animal species—human sentience remains limited to present and change.

Sentience, limits of animal consciousness

We ask ourselves: Who is aware? Who is conscious? Which

---

7. Avolio, C (2025) - *"Schrews Shrink and Regrow Brains, Offering Clues for Human Diseases"*, retrieved from *Neuroscience*, 1 September 2025.

animals have cognition and to what extent do they have it? What varieties of cognition are there? Let's start with the theory of evolution. The first principle is that all organisms descend, with modifications, from remote ancestors. The second principle is that of natural selection, which says that organisms with hereditary variations that adapt better to the environment have more offspring. It is a process of deceptive simplicity but, when applied again and again through aeons, it results in organic systems of exponential complexity and sophistication.

Where should we begin to seek human consciousness? At the molecular level? We know that genes mutate, but following natural selection, gene flow, gene drift and environmental factors. Of course, there are genetic accidents but, in general, there is a change and genes adapt to it. Cultural and behavioural adaptations often precede physiological mutation. To be sure, that process changes the DNA of the offspring, i.e., any changes that take place in the behaviour of the individual generate genetic changes, it does not happen the other way around. What we do during our lifetimes has a great impact on the genes of our descendants.

Evolution presupposes a transition from a non-sentient thing to a sentient being. And, from a sentient being, at some point, evolution deceptively appears to take us to a fully conscious, thinking being.

Here it is necessary to distinguish among several classifications of beings. According to some materialist scientists and philosophers, there is a hierarchy that ascends in order of sophistication: first there are things like plants, which evolve by natural selection, then come beings like sponges and then snails or mice, which also learn from their mistakes during the life of an individual; the next step is beings that learn to select between imagined actions and scenarios, such as elephants and dolphins. Human beings, the most advanced beings, can select from possibilities represented by symbols. So, there is an approach to the issue of consciousness that considers solely an evolutionary transition.

After establishing that nervous processes were a *sine qua non* for consciousness, physicist Erwin Schrödinger (1956) noticed that there was a breakaway point in which consciousness started, even when discussing beings with brains and nervous systems:

"Not every nervous process, nay by no means any cerebral process, is accompanied by consciousness. Many of them are not, even though physiologically and biologically they are very much like the 'conscious' ones, both in frequently consisting of afferent impulses followed by efferent ones, and in their biological significance of regulating and timing reactions partly inside the system, partly towards a changing environment."[8]

How can you tell, then, what determines that difference? Where does consciousness start?

Tantalisingly, Schrödinger provides what he calls a 'tentative' answer to the problem:

"I would summarize my general hypothesis thus: consciousness is associated with the <u>learning</u>* of the living substance; its <u>knowing how</u>* (Können) is unconscious."[9].

He is telling us that what we know without learning is not conscious. A brilliant definition of the difference between sentience and cognition?

---

8. Schrödinger, E (1956) – *What is life?* - *Mind and Matter*, Cambridge University Press, p. 95
9. Schrödinger, E (1956) – *What is life?* - *Mind and Matter*, Cambridge University Press, p. 99. * My underlining.

We have to remember, though, that human consciousness is a combination of sentience and complex cognition.

Most probably, our ancestors, before they acquired language, had a semblance of cognition, a germ of what would exponentially grow into what now is human cognition.

No one doubts that some animal species have basic, limited cognitive skills, especially gregarious species. Some birds, like corvids, and other animals, like elephants, display incredible memory and learning abilities. For instance, they can recognise faces for years. Even hummingbirds recognise specific flowers that they know will provide more nectar. What they do not have is complex cognition or a sense of time. Consequently, they cannot enjoy long-term episodic memory of the type *H. sapiens* has, basically because they do not have recursive language.

What I propose in this book is that human cognition grew from communication, from language. When human communication necessitated complexity, human thought grew with it. The comparison with other animal species, then, has to concentrate on social species.

We know that, until the 1970s, ethology had studied individuals rather than social groups. From then on, ethologists began studying the social structures of animal groups.

Animal behaviour responds to three main sources of input: 1) instinct, which the individual has from birth; 2) learning, which may come from imitation, habituation, etc.; and 3) environment: any changes in the habitat of a group will trigger a response in behaviour.

Oftentimes, animals in social groups learn through teaching by parents or by the group (which is what occurs in terms of human language and infants). This behaviour has been observed in various social species, like whales and crows. Offspring learn their behaviour when parents show them how to obtain food, for instance, by doing something. The parent shows the infant and expects the infant to imitate their behaviour. Sometimes mimicking is spontaneous. More often than not, parents have to go out of their way to show how to hunt or use tools, like twigs, in order to find food that is not otherwise available.

It is quite apparent that some animal species, like other primates, whales or crows, have a degree of oral communication. Teaching takes place when behaviour is shown to offspring. What has not been analysed in other animal species is whether behaviour is transmitted to offspring through communication (explanation) rather than ostension.

Is human consciousness part of an evolutionary continuum?

When we discuss 'learning' in general, we could say that it can be a change in behaviour often grounded on experience. Nineteenth-century thinkers believed that being able to learn by changing behaviour was a criterion of consciousness. That is basically correct. The criterion does not appear to apply exclusively to human consciousness, which is what interests us. Nowadays, human consciousness involves learning through explanation rather than ostension.

That belief changed radically with Skinner's *"behaviourism"*, in the twentieth century. It excluded terms like *"consciousness"* or *"mind"*. Behaviourism—a very limited theory—ended up being falsified. However, it left a legacy: the study of associative learning.

In their book *The Evolution of the Sensitive Soul – Learning and the Origins of Consciousness*, Israeli researchers Simona Ginsburg and Eva Jablonka, both self-confessed materialists and admirers of Dan Dennett, propose a precise link between evolutionism and human consciousness. Ginsburg and Jablonka follow the current philosophical line in terms of research, and apply the taboo against anything that is not materialism:

"We do not, therefore, engage with philosophers who hold dualistic positions with regard to the mind-body problem or with those who do not regard consciousness as the product of biological evolution"[10]

They explore how nature arrives from thing to human mind. They claim their book addresses 'human consciousness' only tangentially:

"It is important to stress here that human consciousness, which laypeople usually associate with the term 'consciousness' and which we discuss from an evolutionary perspective in the last chapter, is not the main topic of this book. Our book is about the origins and evolution of sentience, of minimal animal consciousness — the ability to have basic subjective experiences — rather than the ability to reflect about those subjective experiences, which seems to be the peculiar gift and curse of humans."[11]

When Ginsburg and Jablonka mention *"the ability to reflect about those subjective experiences"*, the mean, of course, cognition. And the quite correctly state that it

"seems to be the peculiar gift and curse of humans". Yes, that is the main difference between human and animal consciousness.

Nevertheless, their book does reach human consciousness

---

10. Ginsburg, S and Jablonka, E (2019) - *The Evolution of the Sensitive Soul – Learning and the Origins of Consciousness*, MIT Press, eBook, p.22.
11. Ginsburg, S and Jablonka, E (2019) - *The Evolution of the Sensitive Soul – Learning and the Origins of Consciousness*, MIT Press, eBook, p.25.

and, in doing so, they repeatedly equate sentience with consciousness *"... evolution of sentience, of minimal animal consciousness"*. They claim that the marker from preconscious to conscious animal is something Ginsburg and Jablonka called "unlimited associative learning" (UAL). As far as I could determine, they provide no explanation as to how the ability is acquired. One of the things that they presumably discovered was that learning depends on how surprising the stimulus is. A totally predictable stimulus does not require learning. According to Ginsburg and Jablonka, *"unlimited associative learning"* is the mark of evolutionary transition from basic 'consciousness'. It requires several elements, among which are the conceptualisation of objects, selective attention and active exclusion, integration through time, spontaneous activity and the existence of a goal, and most importantly: the sense of being an individual separated from others and with a stable perspective over time. Again, no mention that identity (not self-awareness) and time could be logically and exclusively human.

According to their theory, when an animal demonstrates unlimited associative learning, i.e., unrestricted learning, it means that it has the capacity for 'consciousness'. An animal with this characteristic can exhibit complex behaviours and achieve many different goals. An animal that has unlimited associative learning can only achieve it when it is 'conscious'. Two more things that appear in animals with that kind of 'consciousness' are suffering and 'imagination'. 'Imagination' apparently evolves with exploration and learning. Through experience the individual learns to seek what satisfies it and avoid what causes suffering. Learning expands awareness and cognition even further, and the evolutionary cycle continues.

The most basic problem with their approach is that there is no animal with really *'unlimited'* associative learning other than

H. sapiens. That comes from a language that has no limits. Also, the non-human imagination they refer to appears to be short-term, as long-term, voluntary, imagination requires time, and animals—I would argue—do not seem to be capable of strategy or long-term memory.

Animals learn from experience and then habituate. Processes become instinctive. Birds build nests. The classic example is beavers, who learn to build dams. But that is all they can build. Human beings build bridges and dams and skyscrapers. Beavers can only build dams. If they have logs, that is exactly what they would do. They cannot build a bridge. Their capacity to learn is limited.

The same thing happens with primates like chimpanzee Washoe and gorilla Koko, who learnt to use human language with symbols. Yes, they could use language, but their use was limited communicating through linear sequences of words. They could not use language in a recursive manner, and their episodic memory and imagination were limited.

The blurred treatment of sentience and cognition is evident throughout the book. Ginsburg and Jablonka's study is not the exception to the rule. Most current studies of consciousness are grounded on a physicalist approach and ignore any boundary that may exist between sentience and cognition.

In this case, the authors blatantly misinterpret Wittgenstein when they quote the *Tractatus Logico-Philosophicus* and proudly provide an answer (more on that below, on 3.2), which emphasises the biological nature of consciousness, which they equate with sentience:

"[Quote] When the answer cannot be put into words, neither can the question be put into words" [Wittgenstein] ... "How did minimal animal consciousness originate during animal evolution? We argue that this is an answerable question if one can uncover a capacity that is a good marker of the evolutionary transition from

preconscious to conscious animals. This can be, we maintain, an Archimedean point to explore the biological nature of consciousness, of sentience."[12]

∼

Neuroscientists and philosophers may find many arbitrary points along a supposedly 'biological continuum' where cognition (and therefore full human consciousness) appeared in history. The fact remains that the clearest and most elegant way to establish a boundary, to determine where human cognition started, is the onset of complex human language. And that marks the boundary between biological evolution and meta-evolution.

Actually, the ability that marks the commencement of what Ginsburg and Jablonka call "unlimited associative learning" is language. The authors do not appear to be interested in recognising that marker.

∼

In *Mind and Matter* (title which, incidentally, hints at Cartesian dualism), Erwin Schrödinger (1956) puts the matter to rest (pun intended):

"From our own experience, and as regards the higher animals from analogy, consciousness is linked up with certain kinds of events in organized, living matter, namely, with certain nervous functions. How far back or 'down' in the animal kingdom there is still some sort of consciousness, and what it may be like in its early stages, are gratuitous speculations, questions that cannot be answered and which ought to be left to idle dreamers. It is still more gratuitous to

---

12. Ginsburg, S and Jablonka, E (2019) - *The Evolution of the Sensitive Soul – Learning and the Origins of Consciousness*, MIT Press, eBook, p.22.

indulge in thoughts about whether perhaps other events as well, events in inorganic matter, let alone all material events, are in some way or other associated with consciousness. All this is pure fantasy, as irrefutable as it is unprovable, and thus of no value for knowledge."[13].

In one clean sweep, Schrödinger rejects the idea of a continuum from an early animal consciousness, and the newfangled panpsychism of IIT and other theories, as unfalsifiable and *"of no value for knowledge"*.

The linguistic origin of complex cognition

> *"We have found a strange footprint on the shores of the unknown. We have devised profound theories, one after another, to account for its origins. At last, we have succeeded in reconstructing the creature that made the footprint. And lo! It is our own."*
> Space, Time, and Gravitation–
> Sir Arthur Eddington

S. Miyagawa (Massachusetts Institute of Technology) *et al* recently published a study in which they determined, with a high degree of certainty that linguistic capacity was already present in hominins approximately 135 thousand years ago:

"We wish to note the specific role that language may have played in organizing, and hence systematizing, modern behavior. Our proposal is similar to earlier suggestions by Henshilwood and others, but is based on a concrete and verifiable date of approximately 135 kya as the lower boundary for the presence of language. As <u>the most complex communication tool yet devised in nature*</u>, it had a direct and enormous impact on all facets of human life. Language, with its complex system of mental representations and

---

13. Schrödinger, E (1956) – *What is life?* - *Mind and Matter*, Cambridge University Press, p. 93

rules for combining them, is able to create new ways to connect existing symbols and predict new ways of behavior. This is, perhaps, what we see in the time gap between the lower boundary of 135kya for language, and the beginnings of the emergence of rich and normative symbolic behavior starting around 100kya. A way to interpret this gap is that language was central in organizing and systematizing modern human behavior."[14]

Please note that the underlined segment acknowledges, to some degree, the artificial nature of language as *"a tool ... devised in nature"*. Language did not come with the original mammalian package. It was devised by our hominin ancestors as they became human. There must have been some basic degree of animal 'cognition' in them, but, as the linguistic process developed, symbolic thinking must have grown with it.

The paper adds that hominins already had a language template in their brain. I do not subscribe to that view.

"We believe that the time lag implied between the lower boundary of when language was present (135kya) and the emergence of normative modern human behaviors across the population suggests that language itself was the trigger that transformed nonlinguistic early H. sapiens (who nonetheless already possessed "language-ready" brains acquired at the origin of the anatomically distinctive species)* into the symbolically-mediated beings familiar today."[15]

Arguably, hominins had a semblance of cognition which developed with language. The assumption of a *"language-ready brain"* follows a Chomskyan conjecture that leads to the

---

14. Miyagawa, S et al (2025) - *Linguistic capacity was present in the Homo sapiens population 135 thousand years ago*, Frontiers in Psychology - Vol.16, 2025. * My underlining.
15. Miyagawa, S et al (2025) - *Linguistic capacity was present in the Homo sapiens population 135 thousand years ago*, Frontiers in Psychology - Vol.16, 2025. * My underlining.

assumption that language is a monogenetic universal phenomenon. Why would *H. sapiens* individuals (to the exclusion of other hominin species) enjoy a *"language-ready brain"* before developing language? It is quite possible that different languages grew with their different cultures. Cultures are not universal. They are exactly the opposite: the natural result of *H. sapiens* individuals living together in separate groups and developing distinct behaviours and innovations with results in different fields, *inter alia*, defence, growth, kinship, etc.

Miyagawa, and many other linguists, situate the origins of language in the Southern tip of Africa, with the Khoisan family of languages. It so happens that Khoisan languages are 'click' languages, i.e., the have distinct consonantal, phonemic, clicks. That feature of the Khoisan family is quite unique, with some exceptions (Xhosa—a Bantu language—probably copied clicks through proximity, and there is one more, extremely rare, 'click' language in Australia). Other than that, there are no clicks in human languages (approximately 7,000 of them). The lengthy migration process must have seen the creation and extinction of many different cultures and languages, until many became totally unintelligible and separate. The click was lost somewhere during that process. The "language-ready brain" and the universality of language are highly speculative and debatable notions. Assuming a single uninterrupted linguistic descent line is like assuming the linear descent of *H. sapiens*. We are now aware of the fact that the real descent of man took many turns that were neither linear nor quite predictable. In fact, the nature of many of those turns was definitely aleatory.

The authors of the study suggest that normal symbolic behaviour must have commenced approximately 100 thousand years ago:

"Based on the lower boundary of 135 thousand years ago for language, we propose that language may have triggered the wide-

spread appearance of modern human behavior approximately 100 thousand years ago."[16]

An impressive study of language from a neuroscientific perspective conducted by Andrey Vyshedskiy *et al*, confirms that language comprehension can be divided into clusters of abilities. The study had a sample of over 17,000 subjects. At a non-individual, macrocosmic level, those clusters could give us an indication of how language developed historically.

The study classified the clusters, in order of difficulty, as follows:

"The cluster of most-basic abilities, termed 'command-language'-comprehension', included <u>knowing the name, responding to 'No' or 'Stop', and following some commands</u>*. The cluster of intermediate abilities, termed 'modifier-language-comprehension', included <u>understanding color and size modifiers, several modifiers in a sentence, size superlatives, and numbers</u>*. The cluster of most-advanced abilites, termed the 'syntactic-language'comprehension' included <u>understanding of spatial prepositions, verb tenses, flexible syntax, possessive pronouns, explanations about people and situations, simple stories, and elaborate fairytales</u>*"[17].

In terms of cognition, the levels confirm that comprehension of meaning went from understanding basic commands to increasingly abstract thought. The first level—the most basic and direct one—must have included self-awareness and awareness of the environment (basic cognition); the second level—much more human, I would say—some combination of terms, colours, measurement and numbers (already abstract thought); and the third one—with cognition fully

---

16. Miyagawa, S *et al* (2025) - *Linguistic capacity was present in the Homo sapiens population 135 thousand years ago, Frontiers in Psychology* - Vol.16, 2025.

17. Vyshedskiy et al (2024) - *Three mechanisms of language are revealed through cluster analysis of individuals with language deficits, Science of Learning*, Article No.74 (2024) * *My underlining.*

developed—, understanding of space-time, recursive language, and imagination of non-existing entities. Basically, how we went from animal to human in three basic steps. Sounds easy now, but it took many thousands of years.

At an interview, Vyshedskiy stated the view that the studies he conducted on language comprehension have confirmed his Whorfian views concerning language:

"For over 50 years, linguists such as Noam Chomsky and Steven Pinker have proposed the existence of a uniquely human language comprehension mechanism, yet its neurological basis remains largely unknown.".[18]

If language is not monogenetic and universal, individual cultures affect their own languages and the thought processes of the speakers of any given language must be affected by that language.

"Linguists will need to rethink certain aspects of terminology," … *"Since the existence of three distinct language mechanisms was not previously anticipated, current linguistic terminology does not yet accommodate these findings. As Benjamin Lee Whorf observed, language shapes cognition—without precise terms to describe these mechanisms, fully conceptualizing and discussing them remains a challenge."*

"Philosophy faces a similar challenge, as the term imagination is used to describe both involuntary experiences, such as nightmares, and voluntary processes, such as imagining a fairytale. This study provides neurological evidence distinguishing these two forms of imagination. Additionally, the neurological model offers new insights that may reshape philosophical perspectives on human uniqueness.

---

18. Dolan, W (2025) – *Study reveals three distinct mechanisms of language comprehension* – Cognitive Science, March 2, 2025.

Finally, the discovery of the three language mechanisms offers a framework for understanding the evolution of language over the past six million years," ... *"The first mechanism is largely shared with chimpanzees, the second—uniquely human—likely emerged around two million years ago, and the third, which enables full language comprehension, likely developed just 70,000 years ago."*[19]

The studies, both impressive in terms of thoroughness and width, come up with different chronological estimations for the beginning of complex language. Miyagawa suggests 135,000 years, and 100,000 years for full implementation of language, while Vyshedskiy's estimates 70,000. It would not be totally illogical to think in terms of a process of language acquisition with a duration of over sixty thousand years (135,000 − 70,000 = 65,000).

What has been established is that *H. sapiens* chronologically followed other members of the *Homo* genus, which commenced in Africa. Maintaining, however, that the species has a strictly monogenetic origin and that language is therefore monogenetic makes no evolutionary sense. The Chomskyan hypothesis that the human brain has a universal language template may be politically correct, but makes absolutely no evolutionary or scientific sense. Languages took tens or thousands of years of intermingling, copying, extinctions and re-creations. A universal model does not appear to be realistic.

To renowned linguist George Steiner (1975), the universal hypothesis of language makes no sense at all:

"That all men known to man use language in some form, that all

---

19. Dolan, W (2025) – *Study reveals three distinct mechanisms of language comprehension* – Cognitive Science, March 2, 2025.

languages of which we have apprehension are able to name perceived objects or to signify action—these are undoubted truths. But being of the class 'all members of the species require oxygen to sustain life', they do not illuminate, except in the most abstract, formal sense, the actual workings of human speech. These workings, are so diverse, they manifest so bewilderingly complicated a history of centrifugal development, they pose such stubborn questions as to economic and social function, that universalist models are at best irrelevant and at worst misleading. ... Thought is language internalized, and <u>we think and feel as our particular language impels and allows us to do</u>."[20].

The fact that the species is so varied has helped it become more successful. *H. sapiens* could survive in many different environments thanks to its flexibility and its different cultures. The other side of the coin is that isolated populations tend to die off.

There is DNA evidence that *H. Neardenthaliensis* survived, coexisted and interbred with *H. sapiens* for thousands of years after leaving Africa. Many individuals of our species carry Neanderthal and Denisovan genes. Is there reason to believe beyond reasonable doubt that they had no language and that their languages did not influence *H. sapiens'* language, or did not result in the creation of new languages?

Genetic hybridisation, together with cultural and linguistic diversity may have helped the growth of our species and the subsequent exponential development of complex human cognition through adaptive introgression.

Meaning

When our hominin ancestors commenced to talk, they probably did not invent the first phoneme, i.e., a sound that could

---

20. Steiner, G – After Babel, Oxford University Press, (1975), p. 74. * My underlining.

be understood, a sound that carried meaning. Other gregarious species communicate basic meaning, with warnings like 'predator' or 'food'. *H. sapiens* individuals, however, developed a complex combination of phonemes, i.e., they combined sequences of phonemes into words and words into recursive syntax. That gave those phonemes the power to convert basic meaning into complex concepts. The process led to the understanding of abstract thoughts and symbols.

What they did was more than just attaching a mental sticker to an object or an action; they invented sequences and subclauses that eventually resulted in infinite combinations of sounds and more meaning. Meaning became progressively and exponentially complex. There was a moment, then, that was the culmination of an incredibly lengthy process (as stated above, tens of thousands of years). That basic meaning used by those primeval interlocutors, those mental stickers, had become language with complex meaning—ideas that floated between human interlocutors. At that point, language was more than just basic communication; it had been transformed into the communication of complex thought. But it was also something intangible, something infinite and metaphysical that an animal species had created and that made its members human (*H. sapiens*).

When I say that meaning was metaphysical, what I want to state is that meaning had acquired an existence all of its own that resided neither in the mouth or the mind of the individual that uttered the sound nor in the ear or the mind of the individual that deciphered the utterance. That sequence of phonemes that they were sharing was something they had in common, but that did not belong to either of them, or to any other individual member of the culture, or species, for that matter. Those ends that uttered and deciphered meaning were physical indeed, but meaning itself had acquired a life all of its own that from then on would be provided by the collective (by the culture).

Sequences of complex meaning had also become infinite. That fact means that—in all certainty—this paragraph I am typing now has never been written or expressed before. Ever.

That its creation was artificial belies any claims of a purely physical or biological origin of cognition.

It could be said that the moment a hominin attached complex meaning to an utterance and another hominin understood that complex meaning, both had become human beings. That very instant represented the real onset of humanity.

As stated, and to be sure, a more basic kind of meaning existed among animal species before then: grunts, yells and body language to signify 'predator' or 'danger', or 'food', as it happens with other social species, like birds or chimpanzees. The difference between those and the first actual sequence of phonemes resided in how deliberate the vocal/auditory exchange was and the fact that, it could be combined and recombined with other sounds to add more meaning to the message. So, the onset of humanity came about not just because of the creation of the first phoneme, but because it was combined with other sounds in order to produce words, morphology and syntax. Basically, language became recursive and infinite. By recursive I mean that a sentence can be included within another sentence: something like "Andrew said that going ahead with the plan was crazy", where "going ahead with the plan was crazy" is a sentence within a sentence. There is no limit to the possible combinations of sounds and meanings in any given language.

This creation of complex meaning was a unique event which, as stated, took tens of thousands of years to reach its climax. And it was also unique in that it had happened only to our species. Using Nassim Taleb's terminology, I would call it a 'black swan' event, but I would go even further and call it the beginning of meta-evolution. Our species would continue to evolve biologically to some extent, but the main source of variations from that moment on would be cultural and would affect the new hybrid consciousness.

It is worth repeating that although some basic cognition appears to be present in some other animal species, in reality, complex thought would seem to have resulted only from the development of recursive language. It followed language. It is a byproduct of language.

Complex thought grew until it developed into a clear distinction between *H. sapiens* and other animal species. Paradoxically, something that had expanded as a means of communication—i.e., a cultural and social phenomenon—became a process that allowed for a subjective, individual, analysis of reality. Those factors resulted in the possibility of a new witnessing. No creature had witnessed reality before. Not in the same way we witness reality and the universe.

Human witnessing is what we also understand as human consciousness. No other animal species witnesses. No other individual animal is quite an individual as we are. Individuals of other species appear to be only iterations of their species. They may be self-aware, but they have no identity. They live, experience, and are part of nature, nothing else.

H. sapiens is part of nature as well, but not completely. The artificial component of our consciousness allows for a special observation of what we think is 'the rest of nature'. You can call it 'objective reality', 'otherness', 'that which is not me'. But we are the only individuals who, having the two components within our consciousness, can ask about our own experience or that of other individuals. No bat, no duck, no tiger, ponders or questions its own sentience. They just experience. And we know that they just experience because they lack language and they lack complex cognition. They have no access to meaning.

Transmission of cognition through language and culture

What follows probably represents the most compelling evidence for the different nature of sentience and cognition as discrete components of human consciousness. As it happens with many other enigmas—and as stated at the beginning of the book—the truth has been lying, hidden in plain sight. In this case, we are thinking of a skill that has become second nature. But it is not part of what we *are*. It is *something we can learn to do* (if somebody teaches us how to do it). It needs to be taught.

Bipedalism is an ability that has changed our species. It has also changed our appearance and physiology. Between five and six million years ago, when our species split from the common ancestor we had with chimpanzees, our ancestors had four opposing thumbs. Now we have two, and two big toes. Human infants will instinctively walk on two feet (and they are equipped to do it), but if they are taught to crawl or walk on all fours as a permanent form of locomotion, they will do that as well, and they will do it for the rest of their lives.

We often hear that the opposing thumb is one of the great advantages human beings enjoy. Actually, our ancestors had opposing thumbs millions of years ago, before we split from the chimpanzees. What they did not have was toes. Then, when the need arose, the lower digits became atrophied and the physiology of the species changed. In that respect, the preferred advantage was mobility rather than four-limb dexterity. Evolution had to change the shape of the fingers in the lower limbs; they had to become toes. So, bipedalism is a physical ability, a skill that the infants learn instinctively. But the most important skill human offspring have to learn is not a physical one. That ability is language. Culture and thought accompany language.

Language—even if the capacity to learn it is now embedded in our neocortex—still is an acquired skill, and a cultural one at that. Complex cognition is also an ability that has to be learnt. Language—and cognition with it—need to be taught in order for the individual to become fully developed. Human beings need the community to be human.

We can say that language is artificial because it was 'created' by human beings. It does not exist in other species in nature. The same thing happens with complex cognition. We acquire thought, language and knowledge. They are not transmitted genetically.

One interesting piece of evidence to the effect that language is how cognition was introduced into human consciousness is the way the brain uses two types of discourse comprehension: narrative and expository. Language is learnt, at first, through narrative discourse. Toddlers and younger children are told and read stories. They internalise language through what is called the default mode network (DMN), which is normal (the first linguistic skill is listening: they listen and learn). The DMN operates when we think about ourselves and others, and when we use long-term memory and imagination. Language appears to be used at that stage to develop an identity, a sense of self. That is when our first memories appear.

During a second stage of language learning, expository discourse is used. That appears to activate the frontoparietal control network (FPN). Expository comprehension is externally directed. Language is then used to communicate (the second linguistic skill is speaking). The pattern appears to be that language develops thought internally through afference and then language is used, again, to communicate thought (efference).

Patrick House (2022) commences his book *"Nineteen Ways of Looking at Consciousness"* with an amazing observation which he does not pursue to its full extent. Instead, he just comes to a conclusion that chooses to ignore the reality of how full human consciousness is transmitted:

"If I were asked to create, from scratch and under duress, a universal mechanism for passing consciousness from parent to child, I would probably come up with something a bit like grafting a plant. Each parent would donate a small piece of their brain and place it on some sort of growth medium, maybe some agar, or some flour with sugar and yeast, and the child would sort of just expand, like those water-absorbent foam dinosaur toys, into its final shape around the pieces of parental brain until it, too, was conscious. How else could it possibly work?

Instead, something much more remarkable happens in nature. An entirely new creature can grow into a fully conscious version of itself, and the entire process occurs, as if by fiat, anytime a certain kind of single cell with the right mix of nucleotide sugars is kept alive for long enough. Which means that consciousness is not something passed on or recycled—like single molecules of water, which are retained as they move about the earth as ice, water, or dew—from one living creature to the next.

Instead, consciousness can be grown from 'scratch' with only a few well-timed molecular parts and plans laid out. It is not drawn from a recycled tap of special kinds of cells or dredged from a vein of free will. No, the darn thing just grows. From its own rules. All by itself. And we have no idea how or why."[21]

When he states *"An entirely new creature can grow into a fully conscious version of itself, and the entire process occurs, as if by fiat..."*, Patrick House appears to be referring to sentience alone. In reality, the process does not occur *"by fiat"*. House is right, however, when he states that consciousness is *"not something passed on or recycled"*. One component of it—cognition—is not biological, it is not genetic. It involves a linguistically-and-culturally-transmitted process.

Here, what fails is that many scientists still regard consciousness as something that physically—and magically—emerges in the brain, something the brain produces through chemical and electrical reactions. That is how sentience is transmitted in all animal species, including ours. But human beings are not born fully developed. Human beings are born sentient, but not cognitive. That is because sentience is biological; cognition is not.

H. sapiens is what biologists call an altricial species. Our young need an extremely lengthy period to be brought up; that means, they are reared biologically, i.e., mum gives them milk and then they progress to mashed potato or other soft foodstuffs until they grow their teeth; then they are socialised, they play with their parents and other infants; and then they are cognitively educated to be able to live and thrive within human society, which is an artificial environment. Let me repeat this: human society is an artificial environment.

We are not born as cognitive beings. Cognition is artificially

---

21. House, Patrick (2022) - *Nineteen ways of looking at consciousness, Introduction*, Wildfire, pp. 1-2.

added to the biological component of our consciousness. Cognition is transmitted culturally and intersubjectively. Our offspring are born exactly like those of all other animal species. But human children acquire language and culture later, with the intervention of their parents and, sometimes, their group. <u>This has to happen to every individual, every generation</u>. It is difficult to accept, but there is nothing biological or genetic about language or culture.

Every human individual needs to be brought up and educated. The ability to think develops slowly from baby, to toddler, to child, to adolescent, and eventually, to adult. Let us allow the concept to sink in: cognition is so artificial that we are not born with it. It has to be taught every generation. That is the truth that lies hidden in plain sight. That is the truth that scientists and philosophers do not seem, or do not want, to see. That is the wide chasm that lies between materialism and human consciousness.

What happens is that, after countless generations, language has become so ingrained in us that conceiving sentience as something non-linguistic—separate from cognition—, as a discrete layer, appears to be unreal. Language has to come from somewhere, though, but where did it come from?

How has language comprehension and acquisition been explained by the orthodoxy until now? Well, Dronker, Pinker and Damasio (2000)—avowed Chomskyans—claim to ground their 'explanation' on Darwinian evolutionary theory, although the way they state it does not appear as much of an explanation.

"<u>Complex Language Develops Spontaneously in Children</u>

According to Darwin, 'Man has an instinctive tendency to speak, as we see in the babble of our young children; while no child has an instinctive tendency to brew, bake or write.' In the first year of life children work on sounds. They begin to make language-like sounds at 5-7 months, babble in well-formed syllables at 7-8 months, and gibber in sentence-like streams by the first year. In their first few months, they can discriminate speech sounds, including ones that are not used in their parents' language and that their parents do not normally discriminate (for example, Japanese babies can discriminate r and l). By 10 months they discriminate phonemes much as their parents do. This tuning of speech perception to the specific ambient language precedes the first words, so it must be based on the infant performing sophisticated acoustic analyses, rather than on the infant correlating the sounds of words with their meanings."[22]

In the above citation, the materialist bias is quite apparent. There is a fleeting mention of '*ambient*' and '*parents*', but there is no mention of 'culture', or of the fact that children are <u>deliberately taught</u> language. Children cannot learn by themselves. The title of the section *"Complex Language Develops Spontaneously in Children"* sounds like such a basic and simplistic explanation, that I would even qualify it as infantile. And things are evidently not like that at all. We will see that, when children are left to their own devices, they fail to learn any language. Even Darwin's quotation *"Man has an instinctive tendency to speak"* is also wrong in the same way, albeit more understandable as something coming from a nineteenth-century scientist.

Dronker, Pinker and Damasio insist on the existence of a language template in the human brain:

---

22. Dronkers, N F, Pinker, S & Damasio, A (2000) – *Language and the Aphasias*, chapter 59 in *Principles of Neural Science*, Kandel, E R, Schwartz, J H & Jessell, T M, eds., p.1171

"But learning cannot happen without some <u>innate mechanism that does the learning</u>*, and other species exposed to the same input as a child fail to learn at all. In 1959 Chomsky proposed a then-revolutionary hypothesis that children possess innate neural circuitry specifically dedicated to the acquisition of language. The hypothesis is still controversial; some psychologists and linguists believe that the innate capacity for language is merely a general capacity to learn patterns, not a specific system for language, and that the brain areas subserving these skills have no properties that are tailored specifically to the design of language."[23]

The human brain does have the capacity to communicate and to learn language, if taught to do so. It was gradually built into the superior temporal and the inferior frontal circumvolutions or gyri. Yes, human beings can learn all sorts of skills. Claiming that language is innate and universal, however, is equal to claiming that language is biological and natural. The capacity to learn language from the collective, acquired during many generations of hominins that eventually became human, is an ability that was developed—and continues to occur—culturally. Otherwise, children would be cognitive from birth, and use language from birth. Otherwise, cultures would have developed homogeneously. That is not so. Cultures differ in stages of development. Language needs to be taught.

I would not have been able to write this as a five-year-old. This is a fact, and this fact cannot be explained in any other way: cognition is a slowly-acquired—uniquely human—phenomenon. No part of language is innate, other than having organs capable of producing and perceiving sounds, and language centres in the brain that have evolved through

---

23. Dronkers, N F, Pinker, S & Damasio, A (2000) – *Language and the Aphasias*, chapter 59 in *Principles of Neural Science*, Kandel, E R, Schwartz, J H & Jessell, T M, eds., p.1172. *My underlining.

feedback after language was introduced. But learning a language is not an instinct like eating. Sentience is innate, cognition is not. Sentience predates cognition individually and historically.

∼

What does it mean that we are an 'altricial' species? The term altricial comes from Latin *alere*, which is to nurse, to rear, to nourish. It is the opposite of 'precocial'. The young of some species are born well-developed and mobile. Giraffes can run with the herd the same day they are born. They do not need to be taught to walk.

Chimpanzees take about four years to be weaned; they become adults at about thirteen years of age. A human individual takes about twenty years to become a fully operational adult. On average, humans live to seventy years of age. That means that growing up takes almost one third of our lives.

Human individuals are born utterly inadequate. Why is it so? Well, we have to be reared biologically, and then cognitively, that is, we learn to speak, then we are socialised, and then schooled. Learning to speak means that we have to learn a specific language. The language we learn is something artificial created by a particular culture. If not, all human beings would be speaking the same language from the time we were born. When we learn to think, we do so aided by a specific language, which—I repeat—is not universal. Literacy is an important part of our schooling. We learn to read and write. Those are secondary linguistic abilities that come after listening and speaking, which are the more basic ones. Then we learn something even more abstract: numeracy. And then, a mixture of both, algebra.

∼

Failure to socialise and educate a human child results in the incomplete development of the individual. The human being reverts to something similar to its hominin condition. This has happened on several occasions. The absence of human role-modelling can have disastrous results. Through history, there have been several cases of children who grew up surrounded by animals and learned their behaviours. One of the most recent ones is that of a Ukrainian woman, Oxana Malaya, who was born in Kherson Oblast of alcoholic parents. When she was three, her parents left her outside to fend for herself. The toddler looked for warmth and protection among the dogs that were around the house. She was raised by those dogs. She lived in a kennel. All her contact with other living beings occurred there. She literally became one more among a pack of dogs. By the time a government agency rescued her, she no longer behaved like a human. She walked on all fours, and could not speak. She barked and growled and, in general, her behaviour was that of a dog. After rehabilitation, Oxana has a limited vocabulary and her mental capacity is that of a five or six-year-old. Unfortunately, when she required role-modelling, she lost the opportunity to learn how to behave like a human being. We definitely need to be socialised when we are toddlers. *H. sapiens* individuals need language to become fully human, i.e., speaking a language is a *sine qua non* criterion to be human.

If we accept that what we state in the above paragraphs is correct, then we accept that language is not something natural. It does not come with us as part of our biological equipment. The individual process takes years.

We now fear LLMs and AI. We fear the development of lab-grown brains. Paradoxically, being human already is being hybrid. That happened tens of thousands of years ago. *H.*

sapiens, the man who knows, is exactly that, the first hominin who can think, and we can think mainly because of language. When our ancestors 'invented' language—as linguist Daniel Everett would put it—complex meaning came with it.

Writing

What this book maintains is that language is not universal now and that it did not begin as a universal phenomenon; evidence of it is the cultural introduction of writing, which took place less than four thousand years ago, after tens of thousands of years of oral language. The book maintains that the existence of a universal language and of a language template in the human brain, as proposed by Noam Chomsky last century, is an unfalsifiable theory with political implications that have nothing to do with science. Human beings need *to be taught* language. There is no innate linguistic capacity in infants. They cannot learn language by themselves, inductively. Human children are equipped to learn whatever they are taught by their parents or their culture. Language is one of those skills. Literacy is another skill.

Individual languages were transmitted orally for tens thousands of years. Later, different types of writing were created at different times in history and in different places, according to the development of the cultures that needed them. Those innovations involved the notion that communication among human beings did not necessarily have to take place through oral-auditive channels. Oral language is naturally fleeting. When recording of information became necessary, the innovation took place mainly as a mnemonic device for administrative purposes. It was not necessarily just a visual representation of oral language.

One instance of recording information as a mnemonic device might have taken place much earlier than previously known. Time-keeping is almost as old as humanity. We've been measuring time since 'time immemorial', if that makes any sense. A recent article by Bennett Bacon *et al*, *An Upper Palaeolithic Proto-Writing System and Phenological Calendar*[24], published by Cambridge University Press, describes research on paintings, conducted in hundreds of European caves, and on engravings of bones.

The objects of the study were depictions of animals (prey) by *Homo Sapiens* some 37,000 years ago. For a long time, the depictions were believed to be art. The study proposes that they were not art. They were notational mnemonic devices. I would not call them proto-writing, but the study found that frequently occurring signs, like dots, lines and "Ys", paired with figures of animals, were meant to carry meaning. The symbols probably signified lunar months and seasons; they were part of a calendar beginning in spring and recording lunar months. The "Y" sign was indicative of parturition of the particular animal next to the notation.

In any case, the first form of officially recognised writing appeared during the Bronze Age in Sumeria, a conglomerate of city-states in the Middle East, about 3,300 BCE; it was more than just an enduring representation of language. Cuneiform writing, as it is now called, was produced by applying a triangular (wedge-shaped) stylus onto soft clay. Far from being just a visual replication of oral language, it was mostly used to record financial transactions. More than anything, then, it

---

24. Bacon, B et al (2023) - *An Upper Palaeolithic Proto-Writing System and Phenological Calendar*, retrieved from *Cambridge Archaeological Journal* , Volume 33 , Issue 3 , August 2023 , pp. 371 – 389, online paper.

was used as record-keeping. The original purpose of cuneiform writing was then to record numbers; the reproduction of sound does not appear to have been that important.

Several centuries after Sumerian writing appeared, around 2600 BCE, the notational system was adapted to reproduce the phonemes of Akkadian, a language spoken by both, Assyrians and Babylonians. The system persisted for at least two millennia and it was used to record other languages, including Aramaic, a lingua franca imposed by the Assyrians onto the populations they conquered.

The fact that cuneiform had to be adapted to represent language, and that it was used for more than one language, means that writing was originally a cognitive communication device, not just transcription of oral language. Later, it developed into transcription. The writing of other, mainly Asian, languages, like different types of Chinese, and Japanese *kanji*, provide evidence of that development.

Chinese ideograms are used for completely different languages, like Hakka and Hokkien, spoken in Taiwan, for instance, apart from some languages of mainland China, like Mandarin and Cantonese. Ideograms, as the name implies, represent ideas, not sounds. Thus 竹 —which means 'bamboo'—sounds "*take*" in Japanese, "*zuk1*" in Cantonese, "*tik*" in Hokkien, and "*zhú*" in Mandarin. It is clear that the same symbol represents the idea of 'bamboo' but it skips the actual sound of the language. In this case, the representation is directly cognitive, rather than phonemic.

Not all cultures or languages have developed writing; all types of writing involve a second level of comprehension and cultures have advanced at different paces. In the case of writing, the reader must learn to understand symbols. Written languages have their own set of symbols. That is called a script. Reading is basically a form of decoding script.

We have seen that some cultures developed writing in ways that were not phonemic. Japanese has a hybrid system, in which there are ideograms (*kanji*) and two types of syllabaries, called *'hiragana'* (for domestic concepts and sounds) and *'katakana'* (for foreign concepts and sounds).

Historically, different cultures have used writing to address their own needs. Some written systems, as we have seen, were developed in order to track merchandise and commercial transactions; some were used to record history, or to maintain culture and religion (like the *Tanach*, or Hebrew Bible); while others were used specifically for legal purposes or to record historical events.

What is relevant to our hypothesis—that language is cultural rather than universal—is that different cultures have come up with radically different solutions. Oral language is conveyed visually in many different ways.

The materials they used also varied: some used clay, some papyrus; other cultures commenced by inscribing battles on stone stelae. Paper eventually became the standard used everywhere, but especially in the West.

Sometimes, record-keeping skipped language and became solely cognitive. An example of that is the *'quipu'*, a system of cords with knots used in the Inca empire. The *'quipus'* were used by accountants but the way they were used has not been totally deciphered as yet. There was a cast of 'scribes', the *'quipucamayocs'*, or *'quipucamayocuna'*, who were the only ones authorised to make the *'quipus'* and read them. Unfortunately, neither Spanish conquerors nor friars were interested in the way *'quipus'* operated. Their significance was lost to history.

. . .

Time

> *"I am not sure that I exist, actually. I am all the writers that
> I have read, all the people that I have met,
> all the women that I have loved;
> all the cities that I have visited, all my ancestors."*
> Interview, *El País*, 1981-
> Jorge Luis Borges,

Aristotle defined time as *"The measure of change"*. Physicist Carlo Rovelli (2017), an expert on the subject of time, appears to agree with him:

"If nothing changes, time does not pass—because time is our way of situating ourselves in relation to the changing of things: the placing of ourselves in relation to the counting of days".[25]

Yes, but it is important to emphasise that when Rovelli says *'we' 'our way'* and *'ourselves'*, he does not mean all animal species—or all mammalians for that matter—he means exclusively human beings. Because we are the owners of time. We invented time. Aristotle himself implied the same notion when he said *'the measure of...'*. Nobody else measures anything. Rovelli makes that quite clear: *"in relation to the counting of days"*. Only human beings count. Only human beings have numbers. Some human cultures have limited use for numbers.

No other species measures anything; we measure things every day. Science is evidence of that. Science needs to measure in order to explain quantities and qualities. A chimpanzee may compare the length of a stick with the depth of a

---

25. Rovelli, C (2017) – *The Order of Time*, Riverhead Books, p.43

hole. She needs to do that in order to catch ants to eat. But she does not measure using any unit of measurement. We measure objects and land. We measure time in years, months, weeks, days, until we get to microseconds. A crow may need more pebbles in order to perform a task, but he cannot comprehend it in terms of numbers.

Rovelli has questioned the concept of time from many different angles. He says: *"The idea that a well-defined 'now' exists throughout the universe is an illusion, an illegitimate extrapolation of our own experience. It is like the point where the rainbow touches the forest. We think that we can see it—but if we go look for it, it isn't there... The 'present of the universe' is meaningless".*[26]

Much of how we understand reality comes from that *"illegitimate extrapolation"* that Rovelli mentions. We think in terms of time, but time exists only in our thoughts. Rovelli, like all physicists, finds that the concept of time is slippery because a) it is subjective, i.e., it is not a physical object; and b) it is exclusively human and exclusively cognitive.

Sometimes, Rovelli seems to comprehend that *H. sapiens* associate time with language and cognition; but he does not specifically say that time is only a human construct and that it is dependent on language and cognition for its existence:

He says: *"... our grammar is organized around an absolute distinction — 'past/present/future' — that is only partially apt, here in our immediate vicinity. The structure of reality is not the one that this grammar presupposes. We say that an event 'is', or 'has been', or 'will be'. We do not have a grammar adapted to say that an event 'has been' in relation to me but 'is' in relation to you"*[27].

It is quite possible that Rovelli—openly relativist and Einsteinian—has not considered the concept of time as a

---

26. Rovelli, C (2017) – *The Order of Time,* Riverhead Books, p.30
27. Rovelli, C (2017) – *The Order of Time,* Riverhead Books, p.70

subset of cognition. This book suggests time is a byproduct of cognition and thus, of language. Language, as a means of communication, deals with change within society. Language can be relative in certain situations, but it does not need to be relative to express the relativity of time. There are adjectives and pronouns that are relative and that, incidentally, vary according to the way certain cultures perceive space, rather than time. Curiously, Spanish and Japanese have three relative measures of distance. English has only two: "this" and "that". In Spanish there are three: "éste", "ése" and "aquél" to indicate proximity to the subject, proximity to the interlocutor and distance from both. Japanese follows the same principle.

At one point in his book, it becomes apparent that Rovelli is talking about quantum mechanics and specifically about quantum mechanics: *"There is no need in any of this to choose a privileged variable and call it 'time'. What we need, if we want to do science, is a theory that tells us how the variables change with respect to each other. That is to say, how one changes when others change. The fundamental theory of the world must be constructed in this way; it does not need a time variable: it needs to tell us only how the things that we see in the world vary with respect to each other. That is to say, what the relations may be between these variables."*[28]

He keeps on discussing time in terms of relativity and scientific need. But science finds that time is a difficult concept. Time is a device we have invented in order to understand reality at the classical level. Its non-existence is related to life, not science. Eastern thought coincides with the idea of

---

28. Rovelli, C (2017) – *The Order of Time*, Riverhead Books, p.74

dynamic interrelations. Everything is connected within a whole, even change.

Granularity is something characteristic of quantum mechanics. There is a scale, called the "Planck scale," that measures the tiniest chance of time in the gravitational field. The smallest chance of measuring time is called "Planck time." It is $[5.319124 \times]^{10-44}$ of a second. That is, a hundred millionths of a trillion, a trillion of a trillion of a second. A figure more than incomprehensible. Planck's time, then, cannot be measured. No current clock can. As we said, according to physics, if you cannot measure it, you lose a condition for the existence of something. If something is not measurable, it does not exist. On such a minuscule scale, the notion of time ceases to be valid. Time as such does not exist.

Pretty much like Neils Bohr, Rovelli refers to God and provides a clearer idea that he is discussing quantum time when he states that the granularity of time follows a universal characteristic. He probably says it better, perhaps more elegantly, in the original Italian, in "*L'ordine del tempo*":

"*Fiumi di inchiostro versati nei secoli, da Aristotele a Heidegger, per discutere la natura del 'continuo' forse sono stati male spesi. La continuità è solo una tecnica matematica per aprossimare cose a grana molto fine. Il mondo è sotilmente discreto, non è continuo. Il Buon Dio non ha disegnato il mondo con linee continue: lo ha tratteggiato a puntini con mano leggera come faceva Seurat*"[29]

("*Perhaps the rivers of ink that have been spent talking about the nature of the 'continuous' through the centuries, from Aristotle to Heidegger, have been wasted. Continuity is just a mathematical technique to approximate things of very fine granularity. The world is subtly discrete, not continuous. The good Lord did not design the*

---

29. Rovelli, C (2017) – *L'ordine del Tempo*, Adelphi, p.76

world with continuous lines: he did it with a light hand, he sketched it with dots, like a painting by Georges Seurat.")[30]

His explanation of Planck time is, I repeat, of a rare elegance.

But let us go back to the way in which scientists have dealt with time. One of the difficulties many scientists currently find with time resides in their purely materialist approach. There is a connection between time and cognition that science has not been able to pinpoint thus far. And it has not needed to do it maybe because science itself operates within the cognitive field.

In their book *"The Blind Spot"* (2024), Frank, Gleiser and Thomson give an account of time that includes relativity, quantum physics and Eastern thought. They say that, when we think about the passage of time, we cannot have an all-encompassing view of the universe, especially considering the disparate rhythms in space-time:

"This point refutes blind-spot objectivism—the assumption that there can be such a bird's-eye-view (or God's-eye) outside view.

But duration is also one. Measurable times and durational rhythms may differ, but temporal passage itself is immeasurable and resists analysis and explanation. Measured time always presupposes the same ineliminable fact of experienced duration or temporal passage. In philosophical jargon, duration is an example of 'facticity' something that must be accepted but for which no ground or reason can be given. In Buddhist terms, duration exemplifies 'suchness', a

---

30. Rovelli, C (2017) – *The Order of Time,* Translated by Segre, E, Riverhead Books, p.55

concrete character of being for which no conceptual ground can be given."[31].

The authors discuss the passage of time in relative terms. There has to be experience of duration and passage of time. For Buddhism, time appears to be a fundamental phenomenon.

Time—this book maintains—does not exist outside of cognition. That appears logically clear and can be tested. Science has dealt with the consciousness/time relationship, but only marginally. The perception of time is, by its nature, exclusively subjective, or—thanks to language—subjectively shared. Time—as we have already seen—is a clear result of human cognition. It is perceived as change by sentience, but becomes a variable that can be measured only within cognition, e.g., as long-term memory and voluntary imagination.

In his study on language comprehension[32], Andrej Vyshedskiy *et al* confirm that the understanding of time began evolving with the modifier phenotype 70,000 years ago, i.e., that time was first conceived by human beings as they began to understand language.

---

31. Frank, A *et al* – *The Blind Spot* (2024), MIT Press, p.120.
32. Vyshedskiy et al (2024) - *Three mechanisms of language are revealed through cluster analysis of individuals with language deficits, Science of Learning, Article No.74 (2024)*

In the year 401 CE, St. Augustine appears to have been as befuddled by time as current scientists are nowadays:

"What then is time? If no one asks me, I know; if I wish to explain it to one that asketh, I know not: yet I say boldly that I know, that if nothing passed away, time past were not; and if nothing were coming, a time to come were not; and if nothing were, time present were not."[33]

The notion of time is a slippery one. How can we discuss something we experience but we cannot see?

Needing to explain relativity, Einstein combined time with space by conceiving the existence of something he called space-time. The difficulty in understanding that notion is that space and time have different natures as they are subsets of different sets: sentience and cognition. Space is sentient. We can see it. Time is cognitive. We cannot perceive it with the senses.

As stated, Einstein devised the notion of space-time from the perspective of his theories on relativity. He did it in the field of physics and only as part of a scientific analysis. But he went much further: he said that there is not a single time, but an infinity of times. As many times as there are points in the universe. And that notion has consequences in our daily lives. That the clock on the table measures a different time than the clock that is on the floor really does not matter at our daily level. That there are infinite times does, especially if we want to define the nature of time to reach some conclusion, scientific or philosophical. That is to say that, by existing in consciousness, time has expressions as varied as there are individual consciousnesses and as the points in space in

---

33. St Augustine of Hippo (401) – *The Confessions*, eBook, Book XI, p.163 (translated by Bouverie Pusey, E).

which those consciousnesses are, or those that can be imagined.

Another of the coincidences between consciousness and time is the singularity and at the same time the multiple nature of both. Einstein said that there is one time and infinite times. Individually, we think in terms of one consciousness but we also understand there are something like infinite consciousnesses. This has to do with the subjective/objective nature of cognition in Western philosophy, but also with the relativity of consciousness.

Human consciousness does not exist in isolation within the individual brain. It is not an individual phenomenon. What happens in the individual human brain is a physical reflection of a shared human consciousness. As a linguist, I know that consciousness is shared, as language is shared, which is the same thing that happens with time. That does not mean that there is an objective reality. Reality only seems to exist as the sum of the perceptions of individuals and their interactions with others. In terms of theory of mind, we know that the perception of time may vary with age, with certain moments, and we can never be sure other people are perceiving it exactly as we are.

In terms of physics, if something can be measured, if it can be quantified, if it can be defined mathematically, if it's an observable quantity, on which other observable quantities depend, then that entity exists. In that respect, and—in spite of the efforts of some theories of consciousness, like IIT—human consciousness would appear not to exist for physics as it cannot be measured; time, a byproduct of human consciousness, therefore, should not exist either. But we can measure it. Or can we?

We could also say that time is measurable, but only within the classical level. When we get to quantum physics, i.e., to Planck time, what we understand as time ceases to be measurable, it ceases to exist.

If we can say that time does not exist within quantum physics, we can also say that time is a human artifice that we only need in order to understand reality at the classical level. And we could also say that time does not exist in the universe either, except within the cognitive component of our consciousness.

There is a very common phenomenon that occurs relatively often and that could account for the different ways in which we perceive time—as members of the *H. sapiens* species—and how other animal species perceive it. During these situations, especially during accidents, time appears to slow down. Our mind goes into 'fight or flight' mode.

My conjecture is that the neocortex appears to be less active when the individual is under threat. Survival instincts are prioritised over cognitive processes. The allocortex takes over. The conjecture would be that, during those situations, we are sentient but not totally conscious as we normally are. Cognition freezes. Sentience deals with change much faster by itself, and gives us the feeling that things are happening in slow motion. What happens is that the cognitive 'time' we have invented—the one we are used to—actually disappears and is replaced by pure change, which does not give us the opportunity to measure anything. There is a threat to our integrity and we go back to our primeval, natural, being. Our senses are heightened and our short-term memory works overtime and captures details that are normally suppressed by cognition. Our sense of agency becomes more purposeful and we make decisions in microseconds.

Biologist Telmo Pievani (2025) questions how good those and other, instinctive, decisions are:

"This is why the decisions we are making today, whose consequences will affect future generations, are unfortunately not a gift left to us by nature. We have to learn to make decisions through education and culture. <u>Evolution favors the present moment, as survival once depended on seizing immediate opportunities</u>. As a result, we lack true foresight — something many recognize when, for example, they delay starting a diet despite knowing they should.

Homo sapiens is brilliant in calculation, curiosity, and technological innovation, yet profoundly limited in foresight, reasoning, and social judgment. At our core remains that imperfection that Holocaust survivor and writer Primo Levi saw nesting in human nature. People are not beasts, wrote Levi in 'The Drowned and the Saved'; they become so in certain conditions and contexts that reduce them to following their basic instincts. <u>We inherit a dual nature, where culture and experience guide us toward better or worse outcomes, underscoring the need for constant ethical vigilance</u>.*"[34]

Actually, instinctive decisions (hunches) may be—and often are—better than decisions that are pondered over lengthy periods. Our survival often depends on how lucidly we can decide in critical situations, like accidents or threats to our lives, like muggings. We have grown and thrived with both, sentience and cognition. They are both good for us.

Those decisions taken in critical moments are similar to what Eastern philosophy calls 'enlightenment', what Zen, in particular, calls *'satori'*. There are seconds during which we

---

34. Pievanti, T – *Flat Earthers on a Cruise*, The MIT Press Reader, virtual article adapted from the book *Imperfection – A Natural History*, by the same author, 12.08.2025. * My underlining.

'become' the car we are driving, or can dodge the bullet coming towards us. More on that under *Eastern Schools of Thought*, below.

∼

It is indisputable that the Sun rises every day in the East and sets in the West: that happens three hundred and sixty-five times a year. That is, our planet rotates on itself those three hundred and sixty-five times as it circles the Sun.

Apart from being heliocentric, like our planet, and anthropocentric, the concept of time is also cultural. For the ancient Hebrews, the day began when it was possible to distinguish a black thread from a white one. Days ranged from dawn to dusk. The idea had a solid logic, and that is also how animals understand the cycle of day and night. They, and us, follow our internal clocks (circadian rhythms) that regulate our sleep-wake cycles.

Time-keeping is the result of several cognitive processes: understanding cycles longer than days, —which, as stated above, are obvious to other species as well. These longer cycles are seasons, months, weeks. Time-keeping also includes understanding numbers and counting; creating divisions within obvious cycles, i.e., hours, minutes, seconds, etc.

Civilised cultures, like Egyptians and Sumerians had already divided the year into twelve months of thirty days. They also divided the day into twenty-four hours.

The divisions of time appear so natural to us that time seems already instinctive. Let's see how some concepts slowly evolved and how we now assume they are true just because of that.

In the ancient East, in China and Japan, the day was divided into twelve sections of two hours each, which depended on the signs of the zodiac. They varied according to the time of sunrise and sunset.

In Japan, during the Edo period (after the Portuguese and the Spanish tried to introduce Christianity) the day was divided into six parts, between dawn and sunset. The time allocated to those periods varied according to the season, either at night or during the day. The Japanese of those years also measured time according to the bells of castles and temples.

In the West—and now in most of the world—we use a calendar that we inherited from the Romans. The matter becomes quite complicated, because their year lasted ten months. The calendar underwent several changes until it reached the current format.

In 1582 the Gregorian calendar was introduced. Until then the Julian calendar had been used. The difference was that that year, from Thursday, October 4, it was passed to Friday, October 15, with the consequent loss of ten days. That corrected an error that the Julian calendar had had, in which the year was eleven minutes longer than it had to be.

We say that planet Earth is 4.5 billion years old. And we measure the time of its existence in years because it's the only way we can begin to understand that amount of time. That is: a year is the time it takes the Earth to orbit the Sun. But 4,501 billion years ago—if there is anyone who can understand a figure of that magnitude—neither the Earth nor the Sun existed.

Something similar happens with distances measured in light years. A light year is the distance that light travels in a year, that is, 9,460,730,472.581 kilometers—that is, almost 9 and a half billion kilometers. No one can even imagine that distance.

We have learned a lot about time. Time is relative, not absolute. It goes forward. As we have already seen, the universe has an 'arrow' of time that flows towards an entropy, towards an increasing disorder. According to quantum mechanics, when we reach an infinitesimal, quantum level, it seems that time is not necessary. In reality, we would need to understand the way in which time might not exist.

Aristotle said that time consists of two parts, since the present —the now—is not a part. He also said that time is not absolute but relative. The future is going to be at some point, but it still isn't, and the past is no longer there. Therefore, for him, the existence of time was somewhat doubtful.

Aristotle asserted that there was a relationship between time and change, but that they are not the same thing. Time is not change itself. After much rumination, he reached the conclusion that time equals the amount of change. But he also said that it is not a type of measure but a type of order. Then he reached another conclusion: time depends on the soul [the mind], but it can exist without the soul because it can be numbered and numbers are eternal. I do not agree with that last part, as numbers only exist within human consciousness. Numbers are another cognitive and cultural creation. But then Aristotle also said that time can only be counted if there is someone to count it. In the end, we agree: he says that without human consciousness there is no time.

According to physics, if something can be measured, quantified, and defined mathematically, is an observable quantity on which other observable variables depend, then, that something exists.

Let's look at it from another perspective.

We said that there are differences between Einstein's theory of relativity and the discoveries of physics regarding the quantum properties of space and time. There is still no generally accepted theory of quantum gravity, which would link relativity with quantum mechanics.

However, the diversity of opinions regarding the nature of time has been decreasing. Many now understand that the temporal part of the theory of relativity disappears the moment we consider the quantum perspective, that is, from the moment we consider the world at a minute level.

And what has quantum mechanics discovered with respect to time? Well, three fundamental characteristics: its granularity, its indeterminacy, and its relationship with other physical variables.

Although it cannot be measured at the quantum level, time is a measurable, observable and quantifiable quantity at our level, at the classical level. We know, then, that it is real at our level. But there are things we still don't understand. For example, we do not know what the causes of what is called the 'arrow' of time are. We see that time flows forward and not backward; we become aware of the passage of time and are subject to its movement, like all physical objects. Even if the entropy of the system remains constant, increases rapidly, increases slowly, or decreases it by adding energy to the system, the 'arrow' never stops or goes backwards. So, at the classical level, time appears to be real for us, although it may not be fundamental (we'll see: I'll explain it later). At present, physics considers entropy to be a derived quantity and treats time as fundamental. Mathematically, however, if we say that entropy is a fundamental quantity, time becomes an emergent quantity and behaves as such.

The problem lies in how we perceive it and our ignorance of the way it works. Science tells us that, since the hot Big Bang, time progresses in a certain sense for all observers; that, in the same way, that there is an 'arrow of time', that the universe expands according to the laws of thermodynamics, and that entropy grows. Time passes and objects move. Things change. The arrow of thermodynamics and that of time may be related.

Entropy is the extent to which a configuration of particles, once undone, is likely to be able to reconfigure. If you beat an egg, you can't go back to have the whole egg again. Theoretically, it could be done, but the probability is almost zero. In the case of the beaten egg, the probability is infinitesimal. The particles in any system have a finite number of possible configurations.

In physics, currently, a couple of things seem to be certain: the entropy of the universe always grows, and time always goes in the same direction.

Much has been said about the famous experiment that proves that a particle passes through two or more slits at once. That happens when the particle is observed.

When a quantum particle approaches a barrier, it may bounce, or perhaps it will break through the barrier. The evolution of the particle is only determined by observing it. The interpretation of the wavefunction only applies to the unmeasured system. Once the trajectory has been determined (once it has been observed), the behavior of the 'past' is the same as that of the classical level, our non-infinitesimal level.

There is a classical world and a quantum world. What divides one world from the other is the moment when things become defined, when indefinite things become defined in the quantum world. That moment is the present, the now.

Time may or may not be fundamental, and our perception of the 'arrow' of time may or may not be related to thermodynamics. The point is that, at our level, at the classical level, time can be measured, observed and quantified. We can do that because we are cognitive beings.

And then? What do physicists mean when they say that time doesn't exist? Why do they say that? Physics does not include that coffee makers exist, that there are people, that there are buses, but nevertheless accepts that coffee makers, people and buses exist. Physicists accept that coffee makers, for example, 'emerge' at a higher level than physics describes. If time is a fundamental property of the universe, perhaps it also emerges? The thing is, it's easy to understand that a coffee maker is made of particles. But time... what is it made of? Our divisions of time, as we have already seen, are rather arbitrary inventions.

Rovelli says there is nothing mysterious about the absence of time in quantum gravity. What happens is that at the fundamental level there is no variable for time. Time, at that level—as we have already seen, is what they call "Planck time"—so minuscule that it is negligible.

Rovelli says:

"La non-conmutatività determina un ordine, e quindi un germe di

temporalità"[35]. *("Non-commutativity establishes an order, and thus, a germ of temporality").*

According to Rovelli, then, at the quantum level there is a germ of time. Perhaps we could put it in different words. At the quantum level there is change. All sentient species sense change. Time is something useful only to *H. sapiens* at the classical level in order to function culturally.

We have seen, and I have hinted at the beginning of this chapter, that time has preoccupied human beings throughout history, and even during some prehistory, as explained. But that preoccupation does not extend to other sentient animals. Why is that, if we share a degree of consciousness?

The answer lies in human cognition. As individuals, we learn about time from our parents, from the collective, from the culture, from the language they use. Time is important to us as human beings. The way it develops in human minds is individual. Time is taught again and again to every child. It does not come as an instinct. We learn it.

Time is a tool used by the collective to have a clearer idea of when memories occurred and when expectations will occur, if they become a reality. For all we know, our cave ancestors only thought in terms of seasons and the parturition of their prey. That was all they needed.

The assumption is that somebody like Alexander Selkirk—the real Robinson Crusoe—didn't need to measure time in order to survive on his island, except for knowing when days and nights and seasons would occur, and, if he did measure time further than that, it was just as a remnant of his life within human society.

The numbers we need to measure time exist only as far as we

---

35. Rovelli, C (2017) – *L'ordine del Tempo*, Adelphi, p.121

need them. Some cultures and some languages use numbers and time to a very limited extent.

In summary, time appears to exist; but only within human cognition—and as a creation of human cognition—at the classical level. For all other species, and for our species at the quantum level, there is only change.

One last thought about space-time. As discussed above, space-time is a hybrid concept that, by now, has been widely accepted. Space can be perceived by the senses: it is a sentient phenomenon; time cannot be perceived: it is cognitive. It occurred to me that the same thing happens with human consciousness. Sentience perceives; cognition comprehends. It should be easy to understand that it also is a hybrid phenomenon.

A recent article by Nicolae Bochis includes a paragraph that reads:

"Space-time is not just a backdrop where celestial objects are the main players. It's real, dynamic, and it shapes our universe. Imagine it as an invisible construction that holds everything in place. It guides how objects in the universe move and how the events unfold. Without space-time, one could not talk about where or when. Let's explore how Einstein's ideas reshaped our understanding of reality and why they remain some of the most profound ideas in science."[36]

Now let's try to copy Bochis' paragraph replacing the words 'space-time' with 'human consciousness'.

Human consciousness is not just a backdrop where celestial

---

36. Bochis, A – *What Is Space-Time? Breaking Down Einstein's Big Idea* – BC-R, (20.09.2025), extracted from online article.

objects are the main players. It's real, dynamic, and it shapes our universe. Imagine it as an invisible construction that holds everything in place. It guides how objects in the universe move and how the events unfold. Without human consciousness, one could not talk about where or when.

What do you think? Does it work?

Long-term Episodic Memory, Voluntary Imagination, Identity

In his *Confessions*, St Agustine proclaims that there is a dual nature within him and that his consciousness differs from those of other animals; as a man he can think and he can ask, whereas other animals cannot:

"And I turned myself unto myself, and said to myself, "Who art thou?" And I answered, "A man." And behold, in me <u>there present themselves to me soul, and body, one without, the other within</u>.* By which of these ought I to seek my God? I had sought Him in the body from earth to heaven, so far as I could send messengers, the beams of mine eyes. But the better is the inner, for to it as presiding and judging, all the bodily messengers reported the answers of heaven and earth, and all things therein, who said, 'We are not God, but He made us.' These things did my inner man know by the ministry of the outer: <u>I, the inner, knew them; I, the mind, through the senses of my body</u>*. I asked the whole frame of the world about my God; and it answered me, "I am not He, but He made me. Is not this corporeal figure apparent to all whose senses are perfect? why then speaks it not the same to all? <u>Animals small and great see it, but they cannot ask it: because no reason is set over their senses...</u>
*"37

---

37. St Augustine of Hippo (401) – *The Confessions*, eBook, Book X, p.129 (translated by Bouverie Pusey, E). * My underlining.

Augustine describes his thoughts on many occasions. And he thinks about the past, he can remember. When he discusses his memories, he begins by explaining that he thinks in words without a physical sound:

"Nor do I it with words and sounds of the flesh, but with the words of my soul and the cry of the thought which Thy ear knoweth."[38]

And then he explains that God can hear his thoughts.

The Confessions... include long passages in which St Augustine ponders the vastness of memory:

"Great is the power of memory, a fearful thing, O my God, a deep and boundless manifoldness; and this thing is the mind, and this am I myself. What am I then, O my God? What nature am I? A life various and manifold, and exceeding immense. Behold in the plains, and caves, and caverns of my memory, innumerable and innumerably full of innumerable kinds of things, either through images, as all bodies; or by actual presence, as the arts; or by certain notions or impressions, as the affections of the mind, which, <u>even when the mind doth not feel, the memory retaineth, while yet whatsoever is in the memory is also in the mind-</u>* over all these do I run, I fly; I dive on this side and on that, as far as I can, and there is no end. So great is the force of memory, so great the force of life, even in the mortal life of man."[39]

Here, St Augustine, in his subtle way, appears to assign memory to something he calls 'mind', and which I believe we can rightly say is human cognition.

---

38. St Augustine of Hippo (401) – *The Confessions*, eBook, Book X, p.126 (translated by Bouverie Pusey, E).
39. St Augustine of Hippo (401) – *The Confessions*, eBook, p.137 (translated by Bouverie Pusey, E). *My underlining.

Does Augustine discuss imagination? Yes, he does, and in a very interesting context. He first explains that the present has a gooey feeling. That there is a future but it cannot be found. Language uses the future tense to express the possibility of that future.

"The present hath no space. Where then is the time, which we may call long? Is it to come? Of it we do not say, 'it is long'; because it is not yet, so as to be long; but we say, 'it will be long.' When therefore will it be? For if even then, when it is yet to come, it shall not be long (because what can be long, as yet is not), and so it shall then be long, when from future which as yet is not, it shall begin now to be, and have become present, that so there should exist what may be long; then does time present cry out in the words above, that it cannot be long.

And yet, Lord, we perceive intervals of times, and compare them, and say, some are shorter, and others longer."[40]

We imagine that there will be a future. And we can also measure it and compare it. Imagination, much like long-term episodic memory, can explain the time that we have created in our minds that is not the present. What Augustine is saying is proto-Bayesian. We expect the future will occur because our experience indicates that there have been futures before. We know that there are cycles.

At the beginning of *"Transmission of cognition through language and culture"*, I say that something that really proves that the components of human consciousness have different natures is the fact that language and culture are not transmitted biologi-

---

40. St Augustine of Hippo (401) – *The Confessions*, eBook, p.165 (translated by Bouverie Pusey, E).

cally; I also say that the truth has been lying, hidden in plain sight.

Another way to prove that cognitive phenomena are not purely biological—and has been lying, hidden in plain sight—is the fact that children have no long-term episodic memories until they learn to speak.

Sometimes we hear the terms *'childhood amnesia'* or *'infantile amnesia'*. These are easily falsified pseudoscientific explanations for a generalised phenomenon. Bad explanations are easy to change. If you think of how far back you can recall events, you will find that you have no memories—or identity for that matter—before comprehension of language.

One example (and there are many) is an article by Cristina Alberini and Alessio Travaglia (2017) *Infantile Amnesia: A Critical Period of Learning to Learn and Remember*, in the *Journal of Neuroscience*. Alberini and Travaglia begin their article by stating that children forget:

"Infantile amnesia, the inability of adults to recollect early episodic memories* is associated with the rapid forgetting that occurs in childhood. It has been suggested that infantile amnesia is due to the underdevelopment of the infant brain, which would preclude memory consolidation, or to deficits in memory retrieval. Although early memories are inaccessible to adults, early-life events, such as neglect or aversive experiences, can greatly impact adult behavior and may predispose individuals to various psychopathologies* It remains unclear how a brain that rapidly forgets, or is not yet able to form long-term memories, can exert such a long-lasting and important influence. Here, with a particular focus on the hippocampal memory system,* we review the literature and discuss new evidence obtained in rats* that illuminates the paradox of infantile amnesia. We propose that infantile amnesia reflects a devel-

opmental critical period during which the learning system is learning how to learn and remember." [41]

The authors describe *'infantile amnesia'* as the *'inability of adults to recollect early episodic memories'*. The answer to that is that the episodic memories were never formed to start with. Without trying to make a semantic issue out of the term 'amnesia', it is possible to say that there is no such thing as *'infantile amnesia'*. The brain of the infant may have synapses but it has no developed the pathways to the cognitive areas required to create long-term memories. Alberini and Travaglia, themselves, explain that their study has been conducted *"with a particular focus on the hippocampal memory system"* and that *"new evidence"* has been *"obtained in rats"*. Evidence on the workings of long-term episodic memory in the human brain cannot be obtained from rats or any other animal species for that matter. Their study is flawed from the beginning. Long-term memories must form within the neocortical region, as they are exclusively human.

Amnesia is a neuropathological disorder. How can we state that <u>all</u> human infants suffer from the same condition? The answer is easy, they do not suffer from any pathological condition. They do not create memories at a young age because they still have not acquired language. I repeat, long-term episodic memory is a cognitive phenomenon, i.e., it requires language.

Another example, an article by Patricia Lobue (2022), from Rutgers University, retrieved from *Scientific American*, also

---

41. Alberini, C & Travaglia, A - *Infantile Amnesia: A Critical Period of Learning to Learn and Remember*, article in *The Journal of Neuroscience*, June 2017. *My underlining.

demonstrates that the biological nature of cognition is taken for granted. The title of the article is: *"Why You Can't Remember Being Born: A Look at 'Infantile Amnesia'."* The title of the article itself gives a clue that human individuals are supposed to have cognitive abilities from the instant they are born.

Lobue begins by posing hypothetical questions about the beginning of memory and then concludes that infants can form memories:

"In fact, most people can't remember events from the first few years of their lives – a phenomenon researchers have dubbed infantile amnesia. But why can't we remember the things that happened to us when we were infants? Does memory start to work only at a certain age? ... Within the first few days of life, infants can recall their own mother's face and distinguish it from the face of a stranger. A few months later, infants can demonstrate that they remember lots of familiar faces by smiling most at the ones they see most often."[42]

There is no distinction between short-term, or sensorial memories, and long-term episodic memories. Then, the author lists different types of other memories and asks if the problem is whether we cannot form autobiographical memories or if we have no way to retrieve them. According to Lobue, nobody knows. At this point it is evident that the sequence 'long-term episodic memory> language> cognition> prefrontal cortex' is not something being even considered. What is assumed as a fact is that people do experience infantile amnesia:

"In fact, there are lots of different kinds of memories besides those that are autobiographical. There are semantic memories, or memories of facts, like the names for different varieties of apples, or the

---

42. Lobue, V (2022) - *Why You Can't Remember Being Born: A Look at 'Infantile Amnesia' "* – Scientific American, June 10, 2022.

capital of your home state. There are also procedural memories, or memories for how to perform an action, like opening your front door or driving a car... If infants can form memories in their first few months, why don't people remember things from that earliest stage of life? It still isn't clear whether people experience infantile amnesia because we can't form autobiographical memories, or whether we just have no way to retrieve them. No one knows for sure what's going on, but scientists have a few guesses."[43]

Lobue continues by guessing that the sense of identity may have something to do the lack of an 'autobiographical' memory.

"One [factor] is that autobiographical memories require you to have some sense of self. You need to be able to think about your behavior with respect to how it relates to others... Another possible explanation for infantile amnesia is that because infants don't have language until later in the second year of life, they can't form narratives about their own lives that they can later recall. Finally, the hippocampus, which is the region of the brain that's largely responsible for memory, isn't fully developed in the infancy period. Scientists will continue to investigate how each of these factors might contribute to why you can't remember much, if anything, about your life before the age of 2."[44]

Language is briefly mentioned, almost as an afterthought, together with the possibility that the hippocampus is not fully developed. As stated above, no relationship is established between long-term episodic memories, language, cognition and the prefrontal cortex. Memory is taken solely as something that is generated in the hippocampal region, nothing else.

---

43. Lobue, V (2022) - *Why You Can't Remember Being Born: A Look at 'Infantile Amnesia' "* – *Scientific American*, June 10, 2022.
44. Lobue, V (2022) - *Why You Can't Remember Being Born: A Look at 'Infantile Amnesia' "* – *Scientific American*, June 10, 2022.

The last two articles above exemplify the fact that, nowadays, the research community does not support the notion of language as a prerequisite for cognitive abilities, nor does it consider long-term episodic memory or identity as cognitive phenomena, as this book submits. Cognition has become second nature to such an extent that anything related to it is considered as a biological phenomenon.

Carlo Rovelli hints at a philosophical concept of identity within time and appears to think of time as part of a social phenomenon: *"We have shaped an idea of a <u>human being</u>* by interacting with others like ourselves. I believe that our notion of self stems from this, not from introspection. When we think of ourselves as persons, I believe we are applying to ourselves the mental circuits that we have developed to engage with our companions"*[45].

That clearly puts time and identity within human culture, space-time indeed, which would imply language.

And there is more: *"It is memory that solders together the processes, scattered across time, of which we are made. In this sense we exist in time. It is for this reason that I am the same person today as I was yesterday. To understand ourselves means to reflect on time. But to understand time we need to reflect on ourselves"*[46].

Rovelli goes on to explain how we can listen to music in a present that is not a mathematical one, but a sticky, gelatinous reality (here he appears to agree with St Augustine). We live in that present and thanks to that present we can enjoy the glory of Beethoven. He also describes how time and long-

---
45. Rovelli, C (2017) – *The Order of Time*, Riverhead Books, p.105. * My underlining.
46. Rovelli, C (2017) – *The Order of Time*, Riverhead Books, p.107

term memory are totally enmeshed in his identity, not just linked to his present, but actually part of it:

"But there is a third ingredient in the foundation of our identity, and it is probably the essential one—it is the reason this delicate discussion is taking place in a book about time: memory. We are not a collection of independent processes in successive moments. Every moment of our existence is linked by a peculiar triple thread to our past—the most recent and the most distant—by memory. Our present swarms with traces of our past. We are histories of ourselves, narratives. I am not this momentary mass of flesh reclined on the sofa typing the letter a on my laptop; I am my thoughts full of the traces of the phrases that I am writing; I am my mother's caresses, and the serene kindness with which my father calmly guided me; I am my adolescent travels; I am what my reading has deposited in layers in my mind; I am my loves, my moments of despair, my friendships, what I've written, what I've heard; the faces engraved on my memory. I am, above all, the one who a minute ago made a cup of tea for himself. The one who a moment ago typed the word 'memory' into his computer. The one who just composed the sentence that I am now completing. If all this disappeared, would I still exist? I am this long, ongoing novel. My life consists of it."[47]

Time, long-term memory and identity cannot exist without cognition. We would not be ourselves without cognition (and, of course, we would not be ourselves without language).

Ethics, morals, free will

Before beginning to discuss ethics, a comparison is necessary. The comparison needed is one between human beings and individuals of other animal species.

Let us start with a feline: a tiger is a tiger is a tiger. I am sure

---

47. Rovelli, C (2017) – *The Order of Time*, Riverhead Books, p.106

Gertrude Stein would have agreed. An individual tiger is an iteration of tiger. She may sense like an individual, but she has no identity (other tigers do not call her Sandra, although they may recognise her scent); she does not need an identity because her species is not social; she does not venture too far from her territory and —apart from offspring—she does not produce or acquire anything that makes her different from other tigers.

The tiger is inwardly immortal because she has no idea of her own finitude. When she dies, she dies, but she does not ponder death. She does not stress, nor does she need to be competitive, except in a fight, where her life may be at stake. With a limited amount of cognition, she can solve basic problems. She lives in the present. She does not remember being a cub. She does not experience time like a human does; neither does she have long-term episodic memory, or voluntary imagination. She does not plan anything or anticipate anything long-term. Like individuals of many other animal species, she has some feelings and/or emotional memory. But she acts, primarily, out of habit and instinct. She kills the prey she needs to eat—herself or her cubs—and there is nothing wrong with it. And here we come to the crux of ethics. Ethics is exclusively human and cultural. There is no good or evil in nature. The tiger eats the baby zebra because she is carnivorous and she is a predator. There is neither good nor evil in what she does.

Human beings, on the contrary, find that whatever they do has good or bad consequences and has to be considered within that context. Whether something is good or bad is meant to be in terms of humanity as a whole, but it is actually determined by the culture.

Physician David Clawson (2025) shares the opinion that good and evil only occur within the human and cultural context and that they are not as clear cut as they are normally imagined to be:

"The concept of good-versus-evil is itself a sapiocortical construct; it starts with the assembly of symbolic narratives that can evolve into concrete and immovable beliefs. Yet, the idea of good-versus-evil is always relative to the vantage point we hold and the concocted story we tell ourselves and others. To each individual, their immediate construct, narrative, and belief feels solid, fixed, certain, and black or white, but, on observation from multiple vantage points, a much more amorphous, fluid, flexible, nuanced, and shaded reality arises.

Our sapiocortex like no other species' brain (except perhaps AI as an extension of the sapiocortex) can create illusions and hold and fortify faulty constructs, manipulative narratives, deceptive beliefs, and perverse false realities, only to bring about our own agitation, inflammation, catabolism, degeneration, and, finally, destruction. A biotechnological glitch?

The sapiocortex allows us to do and create so many amazing things but also facilitates this unique, opaque, and bizarre ability for predation and destruction of our own species. All the while we seem to be unaware of the false reality we have created. Beyond awareness, the good is lost to the bad; the devil is sneaky that way.

Ultimately, to avoid the blind traps, we will have to discard the false and maladaptive constructs of astral good versus evil, angels and demons, and even the earthly ideas of morality* and immorality, in exchange for a more accurate reality—that of biological safety versus threat. It is threat that shifts phenotype, cells, organisms, and the human brain towards the disconnected, the destructive, and the dark. The physiology of safety shifts phenotype, cells, organisms, and the human brain towards the connected, the creative, and the light. It is not the competing spiritual* powers of good versus evil *nor individual strength of constitution and character that determines these states."*[48]

Morals, from the Latin, *'mores maiorum', (the habits of our elders)* are based on ethics. 'Mores' are what our ancestors did and accepted as correct, and what we, in turn accept as what we are supposed to do. Morals, again, vary dramatically from culture to culture, e.g., what is acceptable in the West may not be acceptable in the Middle East and vice versa. Languages reflect those different biases.

Which brings us to free will which, again, is an issue much more complex than it is thought to be.

According to what I propose in this book, the human brain is divided into two distinct areas that reflect the way consciousness is divided: there is a mammalian brain, and then there is the neocortex, which represents the human development of that brain. The mammalian brain has agency, i.e., any mammal can decide to move or not to move depending on its wishes and/or needs. That means that a human individual can also decide to move an arm, or walk, or scratch his/her head. That possibility of choice, that we share with other mammalians, that basic agency, that generally carries no consequences, is outside the goals of this chapter, or of this point. This point is exclusively focused on free will as an expression of the choice of a human being when it carries consequences, culturally, socially, and in terms of the species. That is, human free will should be taken as a byproduct of the exclusively cognitive, or artificial, component of the individual's consequences, not be confused with the biological one.

---

48. Clawson, D R (2025)– *Technology's Battle of Good versus Evil – Psychology Today* – Retrieved from a virtual article 29.06.2025. * My underlining.

Historically, free will has been discussed for centuries, and it remains being discussed by philosophers and other researchers.

It would be possible to embark on a Byzantine discussion of the pros and cons that we may encounter with respect to free will. To what degree do we have free will? Do we have it at all? There are many topics involved, like how it relates to moral responsibility; the existence of freedom to do otherwise; 'compatibilist' theories of causality or self-determination; 'libertarian' theories of causality or self-determination, etc.

Most free will theories have been historically materialistic, i.e., in ancient Europe it was believed that all things—including souls—were material, and that they were ruled by the laws of nature. This book maintains that human cognition is not natural and it does not abide by the laws of nature. When making a decision, the individual is generally governed by its language and culture, and by other possible circumstances that may influence his/her decision, like timing, predisposition, etc. The responsibility for any action, however, must rest with the adult, fully functioning, human individual, otherwise, our societies would not be able to operate as they do. There may be causal influences on decisions taken by an individual, however, those causal influences are irrelevant once the fully-operational adult individual has taken a decision. The mere notion of human free will is incompatible with determinism. Human society is not deterministic: the results of major events in human history have been almost impossible to predict inasmuch as they were determined by the behaviours of individuals.

Other animal species have limited choices and are mostly governed by instinct, whereas *H. sapiens* has the possibility of individual choice and is only restricted by the laws of nature to a certain degree, e.g., technologically-aided flight is possible, perhaps through co-operation, and sometimes even individually. Basically, and within individual limitations, free will allows us to lead our lives according to our own choices. In human society, the main limitation is moral responsibility, e.g., if I choose to commit an offence or a misdemeanour, I will most probably be facing the consequences of that offence or misdemeanour in a court of law. Moral responsibility is basically grounded on accountability within the culture.

If human societies were deterministic, no advance would have been possible. Nature would have followed its course and human individuals would still be foraging for food. But deterministic views are incompatible with human cognition, and the laws of nature and any events that occurred in the remote past bear no relation to the choices of a human individual in the present.

Philosophers used to argue that nature is causally determined; they deduced that, since human beings are part of nature, they too are causally-driven. I believe human language and cognition (and thus human beings) are not totally natural, hence, I believe the deduction is incorrect.

Proving the existence or non-existence of free will is a futile exercise. We know that free will exists because our human societies function reasonably, i.e., no further evidential support is required for their existence (and therefore, that of free will).

Animals go on lengthy journeys. Swallows fly across continents; whales swim from one extreme to the other extreme of the oceans every year, when they migrate between summer and winter ranges; American bison herds travel thousands of kilometres a year. But they all go from one specific area to another specific area, they have clearly determined routes. They are not adventurous; they travel only to the extent they require to do it. Human beings choose where to go, and when, and why.

Adventurousness, innovation

Human beings are not bound by their instincts. That is how free will operates. Human beings can imagine long-term into the future, consider possible events—positive or negative— and then decide whether travelling and exploring unknown places is safe, or convenient, or worth their while. There was no instinct involved in all of the greatest adventures carried out by human beings, and there were many immense enterprises of that type in history.

During history—recorded or not—as technology allowed them to sail, or travel farther and farther by other means, human beings attempted new exploration feats that seemed impossible before. No other human being had sailed as far South as Bartolomeu Dias, the Portuguese explorer who became the first European to sail around the Cape of Good Hope, in 1488.

According to Maori tradition, the explorers who discovered Aotearoa—later called New Zealand—had come from as far as Hawaii. As they sailed South, they did not know whether they would find any islands. Probably, other expeditions before theirs had failed. When the ancestors of American indigenous peoples travelled across Beringia into the American continent, and further South until they reached Tierra del Fuego, they travelled into the unknown.

The same happened to Christopher Columbus, when he ventured into the unknown across the Atlantic, and to Ferdinand Magellan, when he sailed from Spain towards the Moluccas, looking for spices to season food. Italian linguist and cartographer Antonio Pigafetta (1536) described the five-year ordeal in his book *"Primer viaje alrededor del mundo"*. Magellan never reached the Moluccas, as he was killed by natives in the Philippines. Only eighteen men returned with one ship loaded with cloves, five years after the departure of the expedition (which had left Spain with 260 crewmen on board 5 ships). The captain who successfully commanded the *nao* Victoria on her trip around the world was Sebastián El Cano. The first circumnavigation of the world was a long trip fraught with dangers of all sorts. But it was a massive success for the West. The need was there; the technology was there; and the decision-making was there.

Charles Darwin himself was the naturalist in one of the expeditions that changed the world. Captain FitzRoy had two missions when he commanded the *Beagle* as she sailed from Plymouth: he had to continue charting the coast of South America and he had to come up with a more accurate longitude. Unexpectedly, Darwin conducted an excellent study of the fauna, flora and geology of all the areas visited by the *Beagle*. The trip took five years (1831-1836) and Darwin accomplished much more than what had been expected of him. Based on his observations, he eventually came up with the theory of the evolution of the species.

All of the above exploration trips had several things in common: the explorers did not know whether they would arrive to their destination; they did not know what they would find on the way to that destination; they had a specific goal, or goals, but otherwise, there was uncertainty as to whether they would reach their destination or not. Sometimes they did not even know their destination. They would be exploring to discover what was further afield.

When animals travel, even long distances, there may be some uncertainty, but they always do it by instinct alone because generations of their ancestors have done the same. They never change their destination. Their behaviour is predictable.

Adventurousness and free will go hand in hand. Free will is an exclusively human cognitive ability. There are many possibilities as to why *H. sapiens* decided to travel out of Africa. We may never know how it happened. Maybe they were being chased, maybe they were chasing prey. But travel they did. At some point, somebody decided it was time to leave the continent.

Innovation is something that other animal species definitely lack. Some advanced species may have the intelligence to use a twig as a tool to catch food. Some species may discover that you can break the lid of a bottle of milk with your beak and drink some milk. Those are changes of behaviour. There is no technology involved. Why is that? Innovation is a cognitive skill that requires the implementation of a new idea. The process may be collective or individual. The innovation may be something totally new; or it may involve changes necessary to improve current technology. Innovation may be a different manner of conducting research; it may result in different surgical techniques; it may change the position of rotors in a drone. But successful innovation necessarily results in improvement for the species. That cannot happen to any other species, mainly because the individuals of other species may not be even aware of the fact that they are part of a species.

The result is that humankind has advanced, maybe in a haphazard way, but the improvements are consistent throughout history. What is undeniable is that, in general, individuals of the twenty-first century live better than individuals of the first century CE. Is this the result of a teleological direction in the way the species evolves? The answer to that question falls outside the aims of this book. I can only say that all innovations are the result of human cognitive skills. Within the taxonomy proposed here, improvements in human society are the result of human complex cognition.

Cognition is the origin of many other solely human activities, like drama, literature, and poetry. All of these are possible because of cognition, culture and language. No other animal species has anything remotely similar.

A HISTORICAL PERSPECTIVE

JUDAEO-CHRISTIAN REFERENCES TO CONSCIOUSNESS

The interest in human consciousness is far from new. It was given different names throughout history, but our ancestors were always curious as to why they were different from other animals. Even after millennia of having become cognitive, the similarities with other species must have been sufficient to spur their inquisitiveness.

Towards the end of last century, there was a rise in Scientism (the blind belief in science) especially after Hitchens, Dawkins, Dennett and Harris—the Four Horsemen of Religion, as they were known—castigated religion to the extent they did, and extolled the virtues of science. An agnostic myself, I understand why they did it. There is an evil aura that surrounds organised religion nowadays, and people are abandoning theirs in droves, especially in the West. Hitchens was a highly intelligent and erudite man. I am not so sure about his mates.

Unfortunately, changing one form of belief for another doesn't make any difference or any sense. I'm not trying to excuse organised religion. What I am saying is that, if you look at the big picture, you commence to see a historic pattern: religion and science are extremes of the same paradigm, with philosophy somewhere in the middle. All three are forms of human inquisitiveness. Religion gives you the answer immediately after you ask the question. This is the truth and you have to believe it. Philosophy keeps on asking questions, and science tests those questions and comes up with results. But dismissing any of them to concentrate on the other is a big mistake. Especially when it leads to a blind belief in anything.

From a historical perspective, we can see that different religions have resulted in different philosophical approaches to the eternal questions human beings have always had about nature.

If you are prepared to analyse religion in detail and look for connections through history, it is possible to establish that the Ancient Hebrews—the originators of Christianity and the Judaeo-Christian tradition on which Western philosophy and science are grounded—were proto-Darwinian to a large extent. It is also possible to establish a link between Buddhism and quantum philosophy although, to many, those conclusions may seem farfetched.

As one researches the Book of Genesis, the *Tanach* (or Hebrew Bible), which is roughly the equivalent of the Old Testament of the Christians, it is possible discover many interesting facts. Among other things, one discovers that we, as individuals, may not have souls, but nations do, cultures do. And they may not be immortal, it doesn't matter. The religious part is not that interesting, but Christian dogma and philosophy have influenced the views of the Western world all along, in many transcendental ways.

In recent centuries, there were two important attempts at an exegesis of the Book of Genesis. One of them was by a French theologian called Isaac de La Peyrère who, in the 17th century, came up with a Pre-Adamite hypothesis, i.e., that human beings had existed before Adam. That appeared to explain some of the inconsistencies of the Bible. He seems to have been forced to convert to Catholicism and eventually reneged on his views. Of course, it was the seventeenth century and those things used to happen.

The second attempt was by the famous 20th-century science-fiction author Isaac Asimov. Acknowledging that prior to modern historiography there was no rational version of history better than the Bible, he believed that the Book of Genesis had to be interpreted allegorically and that any literal interpretation of the Bible did not make any sense.

Both of their theories had some merit. I concluded that the Book of Genesis should not be interpreted literally, as that is a major insult to the intellectual capacity of those who wrote it. Like the rest of the *Tanach*, Genesis is mythical in that it is a compilation of oral stories transmitted by the Hebrews around campfires and hearths during countless generations. But it is also allegorical and a recording of what wisdom and knowledge the Hebrews had at the time it was written. Scripture.

It is easy to discover that the Hebrew Bible was written in very difficult times. By the eighth century BCE, the old Kingdom of David and Solomon had been divided into Israel, to the North, and Judea, with Jerusalem as its capital, to the South. Hezekiah was the king at the time. The Assyrians had invaded Israel, which included ten of the twelve Hebrew tribes, and had dispersed its inhabitants. Many refugees from Samaria, Israel's capital, had flocked to Jerusalem, which created all kinds of problems in terms of work, famine, lawlessness, housing, etc. Judea needed a code of law, and Hebrew writing had been recently introduced and was being used on stelae and other official inscriptions. The King decided that it was the right time to produce a history, a religious record, a code of law—something that Hammurabi had done centuries before—but with the authority of the God of the Jews behind it. The Bible was born in times of crisis, but it was meticulously written, as one of its purposes was to be used as the Law of the Nation.

Interpreted literally, the Book of Genesis does not make a lot of sense these days. But two things are important: it has to be interpreted correctly, and the reader needs to consider the time in which it was written and make allowances for that.

Adam was made from dust. Asimov says:

"The Biblical writers knew nothing of microscopic life, but dust is not a bad way of describing it, in the absence of knowledge. Microorganisms are as small as dust grains, after all."[1].

Yahweh was not meant to be a potter, but is there a better way of explaining the origin of human life to a nation of illiterate shepherds?

As I said before, the literal version of what happened in the Garden of Eden—what most Christians believe— doesn't make a lot of sense. Let's see: God created Adam from dust. Then he created Eve from his rib. They were the first human beings. He put them in the Garden of Eden. He ordered them not to eat from a tree. He said if they did, they would surely

---

1. Asimov, I – *In the Beginning*, Open Road, Integrated Media, p.86.

die. Then the Devil, disguised as a snake, tempted Eve and convinced her that she should eat. Eve disobeyed God and persuaded Adam to do that as well. They both ate 'an apple'. They discovered they were naked and probably had sex, although that is not quite clear. God, in a fit of rage (?) expelled them from the Garden of Eden. He told Adam he would have to work and that he would die and turn to dust. He told Eve she would hate snakes and would give birth with pain. But then, after they ate the 'apple', they did not die, even when God said they would. He was wrong. God was wrong? Later they had two sons, Cain and Abel. Cain killed Abel. God became very angry (again?) And sent him away. He went away worrying that somebody would kill him. Who, if there was nobody else on Earth? Literally. He went East of Eden and built a city in the Land of Nod. Who were the inhabitants, if there was nobody else on Earth? He came back with a wife. Who, if there was nobody else on Earth? … and so on and so forth.

The literal Christian interpretation that Adam and Eve were the only human beings on the planet does not appear to hold water, or to make a lot of sense from a biblical perspective. Actually, it does not make any sense at all. Saint Paul said they had been created immortal and became mortal afterwards. Did he mean it literally? If he did, it does not make a lot of sense.

I would say it is quite probable that the scribes of King Hezekiah created a very beautiful, allegorical version of the myth. If we believe that Genesis should be understood allegorically, all of the inconsistencies of the Bible disappear. Of course, the presence of a deity in the picture is not quite what many people in the West, especially scientists and philosophers, would accept nowadays. But we have to take into account that the Bible was written almost three thousand years ago, and our understanding of reality has, to a large extent, changed.

Let us respect the ancient Hebrews as an intelligent nation. As stated above, I would guess they were proto-Darwinian; they had some idea, some inkling, that human beings had evolved from other, more primitive, species. Being shepherds, they had some knowledge that species could be improved for certain purposes by means of breeding. They knew about horses, donkeys, and mules. They probably knew about other primates.

We have seen that Aristotle and Plato, by using the word *zoon* when referring to human beings, appear to have implied that human beings were considered primates. They also used *"animal with thought/language"* (ζῶον λόγον ἔχον, *zoon logon echon*). It is quite possible that that was the natural understanding in the days of the *Torah* and before, when all the biblical stories originated: that it was still understood that human beings had developed from apes, that they thought we were descended from lower animals.

Did the Hebrews believe this during Hezekiah's time, when at least part of the Bible was compiled? Or had the traditions and stories already lost their original meaning? If we put the question into chronological perspective, Hezekiah lived in the 8th century BCE and Aristotle lived in the 4th century BCE. Even allowing for the fact that they were located in two different geographic areas, the fact that there was a gap of four centuries between them, and that Aristotle still considered himself a primate gives me a clear indication that people in Hezekiah's time were aware of our animal origins, i.e., pretty much proto-Darwinian in their ideas.

Judah was becoming slowly urbanised, but they probably understood those traditions and legends with their original meaning. We have to remember that this was the early Iron Age and that living in caves, for instance, was not strange.

The majority of human beings were not city dwellers, even in the days of Jesus Christ. Many, like John the Baptist and his mother, or before then, Lot and his daughters, lived in caves in the desert. And of course, for many generations, Hebrews lived in tents like some Bedouin still do.

There are stories in the Bible that may allegorically refer to different stages of human evolution. Isaac, for instance, was tricked into giving his inheritance to Jacob, his youngest son. Isaac was blind and Jacob made him believe he was Esau, the eldest, by covering his hands with goatskin. Apparently, Esau, a hunter, was extremely hirsute. When Isaac touched Jacob's hands, he thought Jacob was Esau. The strange part of this old story is that Esau was so hirsute that his hands felt like the skins of goats. Allegory or exaggeration? The story refers to the period of the early settlement of Canaan. It has several readings, both literal and allegorical. In the days of Abraham and Isaac, towns were not the norm in Canaan, many people still lived in the open, in caves, or were nomads. The question is: When the Bible was written, did they think that humans had always been developed and separate from other species? Or perhaps that there were some humans that were less developed than others? Maybe the story makes reference to the Edomites, Esau's descendants, who lived in current day Jordan, where Petra is. The *Torah* does not regard them as very advanced. The thing is: if it was reasonable to think that a human could be as hairy as a goat, the mental gap between human and animal was not as wide as it was to become later.

Stephanie Moser, an iconography specialist, attempts an explanation:

"Another biblical figure who conveyed a sense of the distant past was Esau, the brother of Jacob. Esau is described as being hairy all over and is frequently depicted as a wildman. The description of his physical appearance as hairy relates to his status as a hunter who lived in the wild. It also relates to his being the forefather of the Edomite nation, who may have been perceived by the Hebrews as hunters."[2]

What Moser does not explain quite well is why hunters were supposed to be hirsute. Was there the impression that there were humans who were at different stages of evolution? The Hebrews, of course, did not understand concepts like genus and species. Maybe I am reading too much here, but it is possible.

As I state above, nowadays we know that in Europe, tens of thousands of years ago, *H. sapiens* individuals coexisted and interbred with individuals of the *H. Neanderthaliensis* species. There were other closely-related, apes that coexisted with humans and were less advanced, like the *Denisovans*. Is it possible that, in the Middle East, the knowledge of those species had survived until biblical times? Again, it is possible.

Moser confirms that at a later stage, in classical times, barbarians were depicted as primitive, i.e., that outsiders or enemies were vilified that way:

"It was thus, at this early stage, that key icons for signifying the distant past were established, including the club, the animal skin, nakedness, hairiness and dark skin colour. These attributes effectively became visual symbols that played a critical role in communicating primitiveness and in separating non-Greeks and non-

---

2. Moser, S – *Ancestral Images: The Iconography of Human Origins*, Cornell U.P. (1998), p.43.

*Romans. They signified an outsider or barbarian status and summed up the qualities of the non-civilized existence."*³

So, even when this did not depict Romans or Greeks, the idea that there were humans who were not as evolved as others was not that strange. Through St Paul, Christians may have inherited the concept, but understood it as a difference between Christians and pagans. Moser explains:

*"In a general sense the visual icons developed in early Christian, medieval and Renaissance times functioned as part of a wider dialogue on how non-Christians were to be defined. This dialogue was inherently visual and relied on symbolic ways of conveying the primitiveness of a pagan existence."*⁴

Early Christians obviously never gave a second thought to the fact that, if there were human beings who were not very advanced or not very civilised in the Bible, it was possible that the ancestors of the whole of humanity were not very advanced or very civilised either. The Ancient Hebrews, however, appear to have had the intuition that humans were descended from apes.

Let us say the myth of Adam and Eve deals with the beginning of human consciousness. Then, it all makes sense. But we have to remember that a myth is a traditional story that attempts an explanation of distant origins. It attempts an answer to the ever-present human questions: Where did we come from? Who are we? Why are we different? Are we animals? Are we angels?

---

3. Moser, S – *Ancestral Images: The Iconography of Human Origins*, Cornell U.P. (1998), pp.169-170.
4. Moser, S – *Ancestral Images: The Iconography of Human Origins*, Cornell U.P. (1998), p.169.

Imagine Adam and Eve as a couple of hominins who had probably walked away from the rest of their group, or clan, and had started communicating more complex thoughts. Of course, it's an imaginary scene. In real life, as we know, the beginning of human consciousness took tens of thousands of years.

Viewing it as an allegory, Adam and Eve, as the hominins they were, would have known nothing about their future and their eventual death. In that sense, they would have been immortal, indeed. They would have lived day-by-day. But they were not the only hominins on Earth. They were the first ones who understood each other; who thought; who could decide whether their actions would be good or bad for their culture. No other animal species had anything similar to a concept of good or evil.

In that sense—if we can put ourselves in the position of a person who lived in those days, if we could think that such a couple ever existed—they could have been, indeed, the first humans. The fact remains that there are many details in the Book of Genesis that point towards an allegory.

After Cain killed Abel, he left his parents and went to the land of Nod, East of Eden. He was scared anybody could kill him. The literal interpretation has no explanation for that. He found a wife, and a son, Enoch, and founded a city, which he called after his son. Had Adam and Eve been really the only human beings on Earth, there would be no explanation for any of that. In the allegory, Cain's partner had probably learnt language, and they would have also taught their children to speak and think. In the allegory, humanity commenced to grow and developed as a species.

Myths normally include supernatural beings. In this case, there is a deity, which is what Hebrews believed. But if the Book of Genesis is interpreted allegorically, it makes a lot of historical, Darwinian, sense. The opposite is true of the literal, religious, interpretation of it.

Almost eight centuries later, Saul of Tarsus—aka St Paul—a Jew from the Greek diaspora, born in the Roman province of Cilicia (now Turkey), added some important details to the myth. As opposed to the rest of the followers of Jesus, who were illiterate, Paul could read and write and he was fluent in Greek and Latin, as well as Hebrew.

As mentioned before, Paul's interpretation of the myth was that before the Garden of Eden, Adam and Eve had been immortal. Of course, they were not yet human. They had no idea of their own finitude.

Paul—who was probably familiar with the Greek philosophers (especially Plato)— interpreted the myth with the new mentality of somebody coetaneous with Jesus, but with a much more sophisticated level of education. According to Paul, what happened at the Garden of Eden was that Adam and Eve had received (a God-given) immortal soul (*psyche*). A suitable explanation for human consciousness at the time. The idea of the 'immortal' soul is, of course, Platonic.

In the *Tanach*, the Hebrew Bible, the word most used to signify consciousness is *nephesh* (Hebrew נפש "breath of life"), i.e., the senses, or sentience. The word appears seven hundred and fifty times in the Tanach.

The most important vocabulary addition of the New Testament is the word *psyche* (Greek: ψυχή, or "soul"), which is actually used one hundred and five times and which, I would say, was meant as human consciousness with the inclusion of its second layer: cognition.

In ancient times, people asked themselves whether the soul humans had was different from the rest of the other animals because they knew we had a special kind of consciousness. Well, we do. We have human consciousness, and it is different from those of other species.

Some Greek philosophers (e.g., Plato), in a most civilised fashion, believed humans had an immortal soul. As stated, St Paul introduced an individual immortal soul based on Plato's but added details of his own to fit his Christian doctrine. Humans are better than and separate from all other animals. The Sadducees, who were the most important Jewish faction, rejected the idea of the immortal soul then and Jews have rejected it ever since. Christians went their own separate way.

Aristotle thought we had souls, but mortal ones, like all other animals. Much later, St Thomas Aquinas went one step further and specified that only humans have immortal souls. He said that animals have souls that are mortal. These are the sort of decisions that plague Christian theology. However, we had become separated from our fellow creatures in the animal kingdom at a much earlier stage. St Thomas Aquinas, in that respect followed what had been said before him. And every time you make a decision like that you have to explain the whys and the wherefores, and that is not always an easy task. What follows never, never, makes a lot of sense.

If we depart from the fact that animals are alive—that they have a *nephesh*, according to the Hebrews, i.e., that they are sentient—, then what Adam and Eve acquired in the Garden of Eden was what the Greeks called *psyche,* what St Paul imagined as the individual immortal soul, and what we now call cognition. If properly interpreted, then, the myth explains the acquisition of the two layers of human consciousness.

. . .

Ludwig Wittgenstein

"What is your aim in philosophy? - To show the fly the way out of the fly-bottle."
- Ludwig Wittgenstein

"I don't wanta hear all your word descriptions of words words words you made up all winter, man I wanta be enlightened by actions."

Jack Kerouac - *The Dharma Bums*

"I went to the woods because I wished to live deliberately, to front only the essential facts of life, and see if I could not learn what it had to teach."
Henry David Thoreau - *Walden*

Gautama had evidently guessed the dual nature of human consciousness. Buddhism is based on that taxonomy. Not many other people in history did. Ludwig Wittgenstein was probably one of them. I do not know to what extent Wittgenstein saw cognition as an artificial, meta-evolutionary, addition to human consciousness. He did not state it in so many words, but his writings, especially his *Tractatus Logico-Philosophicus*, indicate a profound understanding of the issue. The counterpart is that many philosophers—notably his teacher and mentor, Bertrand Russell—guessed they were before a brilliant intellect, but failed miserably to understand his early message.

That Wittgenstein's theories remain incomprehensible to many is evidenced by the fact that now—almost a century later—scientists and philosophers are still trying to understand, explain and measure human experience.

∽

Wittgenstein was the scion of a very wealthy Austrian family. His father, Karl Wittgenstein, was one of the most powerful industrialists of his time. Ludwig had four older brothers and two sisters who were also bright in different ways, but by the time he had reached his twenties, it was evident that he was a uniquely gifted individual. In fact, he was a polymath.

Following in his father's steps, he started studying mechanical engineering in Berlin. By 1908, he was training in aeronautics at Manchester University. While conducting his research in aerodynamics he invented a different propeller. In order to solve the problems of his design, he studied the mathematics involved. One of the pioneers in mathematical logic at the time was Bertrand Russell, so Wittgenstein applied to study under Russell's guidance. Within two years, Wittgenstein had nothing to learn in that field and was arguing with Russell about his theories. He soon abandoned engineering and devoted all his energies to the study of philosophy. He discovered that all of the great philosophers had made *"disgusting mistakes"*.

Wittgenstein returned to Vienna and, at the outbreak of WWI, volunteered as a private in the Austrian Army. Eventually he was given a commission and sent to the Italian front. Towards the end of the war, he was taken prisoner by the Italians and confined in Monte Cassino. During the intervening years he had been writing the *Tractatus Logico-Philosophicus*, the manuscript of which he had kept with him while he was a prisoner of war.

Wittgenstein was not interested in debates. He expected to be clearly and completely understood. Otherwise, there would be no point in saying anything at all. Bertrand Russell once told him that, rather than just stating what he thought was true, he should provide arguments. Wittgenstein replied that providing arguments would spoil the beauty of the idea. The *Tractatus* is based on that kind of logic. It is either clear, or it is unclear. I believe it has been mostly the second.

From Russell's *Introduction* to the *Tractatus*, it becomes apparent that he had not quite understood what Wittgenstein was saying. The book is important—he says—and it is not incorrect, but he cannot explain why:

"Such an hypothesis is very difficult, and I can see objections to it which at the moment I do not know how to answer. Yet I do not see how any easier hypothesis can escape from Mr Wittgenstein's conclusions. Even if this very difficult hypothesis should prove tenable, it would leave untouched a very large part of Mr Wittgenstein's theory, though possibly not the part upon which he himself would wish to lay most stress. As one with a long experience of the difficulties of logic and of the deceptiveness of theories which seem irrefutable, I find myself unable to be sure of the rightness of a theory, merely on the ground that I cannot see any point on which it is wrong. But to have constructed a theory of logic which is not at any point obviously wrong is to have achieved a work of extraordinary difficulty and importance. This merit, in my opinion, belongs to Mr Wittgenstein's book, and makes it one which no serious philosopher can afford to neglect."[5].

Knowing Wittgenstein personally, understanding his genius, but unable—or unwilling—to say that he had not understood, Russell stated:

"There are some respects, in which, as it seems to me, Mr Wittgen-

---

5. Wittgenstein, L (1922) – *Tractatus Logico-Philosophicus*, Kegan Paul, Trench, Trubner & Co., Bertrand Russell's *Introduction*, p.19.

stein's theory stands in need of greater technical development"[6]. [Please, Ludwig, explain what you're saying!!]

Russell was revered as one of the most important logicians and thinkers of his time (and to some extent, he still is). To admit that he had not understood what is probably the ultimate theory within his discipline was courageous and pathetic at the same time. He was the first one, but he would not be the last one. More than a century has passed and it is quite obvious that Wittgenstein remains misunderstood and ignored by the majority of philosophers, logicians and scientists.

Wittgenstein was quite clear in that he did not care. He wrote for his peers:

"This book will perhaps only be understood by those who have themselves already thought the thoughts which are expressed in it — or similar thoughts. It is therefore not a text-book. Its object would be attained if there were one person who read it with understanding and to whom it afforded pleasure.

The book deals with the problems of philosophy and shows, as I believe that the method of formulating these problems rests on the misunderstanding of the logic of our language. Its whole meaning could be summed up somewhat as follows: What can be said at all can be said clearly; and whereof we cannot speak thereof one must be silent."[7]

Trying to explain sentience, Wittgenstein says, is an exercise

---

6. Wittgenstein, L (1922) – *Tractatus Logico-Philosophicus*, Kegan Paul, Trench, Trubner & Co., Bertrand Russell's *Introduction*, p.17.
7. Wittgenstein, L (1922) – *Tractatus Logico-Philosophicus*, Kegan Paul, Trench, Trubner & Co., Preface, p.23.

in futility. There is no other way of understanding him. Language is not equipped to do that.

The philosopher clearly states that the idea of space-time is a human construct:

"6.3611 - We cannot compare any process with the 'passage of time'—there is no such thing—but only with another process (say with the movement of a chronometer).

Hence the description of the temporal sequence of events is only possible if we support ourselves on another process.

It is exactly analogous for space. When, for example, we say that neither of two events (which mutually exclude one another) can occur, because there is 'no cause' why the 'one' should occur rather than the other, it is really a matter of our being unable to describe one of the two events unless there is some sort of asymmetry. And if there 'is' such an asymmetry, we can regard this as the 'cause' of the occurrence of the one and of the non-occurrence of the other."[8]

In this case, the movement of a chronometer is a product of cognition. Time can only be understood in human terms. There is no other way. The same thing happens with space. He is discussing relativity and causality. Wittgenstein can see that there is a parallel between reality and our comprehension of it.

But he goes even further and rejects causality again, as well as Bayesian statistics:

"6.36311 - That the sun will rise to-morrow is an hypothesis; and that means that we do not 'know' whether it will rise."[9]

---

8. Wittgenstein, L (1922) – *Tractatus Logico-Philosophicus*, Kegan Paul, Trench, Trubner & Co., p.86.
9. Wittgenstein, L (1922) – *Tractatus Logico-Philosophicus*, Kegan Paul, Trench, Trubner & Co., p.87.

"6.37 - A necessity for one thing to happen because another has happened does not exist. There is only 'logical' necessity."[10]

"6.371 - At the basis of the whole modern view of the world lies the illusion that the so-called laws of nature are the explanations of natural phenomena.".[11]

Evidently the phrase [Newtonian] 'Laws of Nature' is a misnomer. Nature has no laws. Scientists have invented principles that explain certain phenomena or events in nature. They are only human explanations of those events that happen and we need to understand.

Without even mentioning the dual nature of human consciousness, Wittgenstein constantly reflects on the impossibility of an answer:

"6.52 - We feel that even if 'all possible' scientific questions be answered, the problems of life have still not been touched at all. Of course there is no question left, and just this is the answer."[12].

Nature needs no explanation. It just is. Human beings— after developing cognition—need to understand, hence, religion, philosophy and science.

The paradox is that in order to understand we have to explain that there is something inexplicable:

"6.54 - My propositions are elucidatory in this way: he who understands me finally recognizes them as senseless, when he has climbed out through them, on them, over them. (He must so to speak throw away the ladder, after he has climbed up on it.)

---

10. Wittgenstein, L (1922) – *Tractatus Logico-Philosophicus*, Kegan Paul, Trench, Trubner & Co., p.87.
11. Wittgenstein, L (1922) – *Tractatus Logico-Philosophicus*, Kegan Paul, Trench, Trubner & Co., p.87.
12. Wittgenstein, L (1922) – *Tractatus Logico-Philosophicus*, Kegan Paul, Trench, Trubner & Co., p.89.

He must surmount these propositions; then he sees the world rightly"[13].

It is said that Wittgenstein reneged on his earlier ideas, on his *Tractatus*. That is not so. *Philosophical Investigations* does not reject any idea in the *Tractatus*.

I could go on and on quoting Wittgenstein from the *Tractatus* or from other sources. Yes, there are some inconsistencies here and there, but the only conclusion we can reach after having read him is that what he wanted to state was totally ineffable. He did it, though. He discussed human consciousness, culture, time, and different views of the world, and then he did what only someone like him could do: he told us to throw away the ladder and stand on thin air. That is what I am trying to do here.

Arthur Schopenhauer

Few Western thinkers were impacted by the profoundness of Eastern thought. He was the first important Western philosopher to mention the Upanishads (and probably the second, after Leibniz, to acknowledge the importance of Eastern thought). Although he was deeply influenced by Immanuel Kant's idealism, Schopenhauer's thought and his connection with Eastern mysticism influenced, in turn, thinkers like Ludwig Wittgenstein, and others, who recognised the important difference between sentience and cognition in human consciousness.

Schopenhauer used his own terminology, which sometimes makes his thinking difficult to follow in important works like

---

13. Wittgenstein, L (1922) – *Tractatus Logico-Philosophicus*, Kegan Paul, Trench, Trubner & Co., p.90.

The World as Will and Idea. That minor obstacle disappears once you become accustomed to his terminology. The importance, depth and width of his contribution to Western philosophy cannot be underestimated.

The message is quite clear: he based his philosophy on Kant's, and recognised idealism in general, but had a strong Eastern feeling in much of his thinking, as he acknowledges early in his work:

"The philosophy of Kant, then, is the only philosophy which a thorough acquaintance is directly presupposed in what I have to say here. But if, besides this, the reader has lingered in the school of the divine Plato, he will be so much the better prepared to hear me, and susceptible to what I say. And if, indeed, in addition to this is a partaker of the benefit conferred by the Vedas, the access to which, opened to us through the Upanishads, is in my eyes the greatest advantage which this still young century enjoys over previous ones, because I believe that the influence of the Sanscrit literature will penetrate not less deeply than did the revival of Greek literature in the fifteenth century if, I say, the reader has also already received and assimilated the sacred, primitive Indian wisdom, then is he best of all prepared to hear what I have to say to him."[14].

But Schopenhauer was preoccupied with sentience, with how it is that we have perception. Kant left perception as a fundamental, something that comes from the outside. Afference.

The Western side of Schopenhauer's idealism is nothing new:

"All that in any way belongs or can belong to the world is inevitably thus conditioned through the subject, and exists only for the subject. The world is idea.

This truth is by no means new. It was implicitly involved in the

---

14. Schopenhauer, A (1909) – *The World As Will and Idea*, Vol. 1, Kegan Paul, Trench Trübner & Co., pp.8-9.

sceptical reflections from which Descartes started. Berkeley, however, was the first who distinctly enunciated it and by this he has rendered a permanent service to philosophy, even though the rest of his teaching should not endure."[15].

He then points to the early recognition of this truth as a fundamental tenet of Vedānta philosophy. Schopenhauer reflects on both layers of human consciousness, and also on the wholeness of perception, with a strong Eastern influence:

"So then the world as idea, the only aspect in which we consider it at present, has two fundamental, necessary, and inseparable halves. The one half is the object, the forms of which are space and time, and through these multiplicity. The other half is the subject, which is not in space and time, for it is present, entire and undivided, in every percipient being. So that any percipient being, with the object, constitutes the whole world as idea just as fully as the existing millions could do; but if this one were to disappear, then the whole world as idea would cease to be. These halves are therefore inseparable even for thought, for each of the two has meaning and existence only through and for the other, each appears with the other and vanishes with it"[16].

The Eastern influence is noticeable in the way he analyses the dynamic aspect of perception, but also on the interconnectedness of the process. The whole perspective aligns with quantum physics. If the observer is not there, the process is incomplete. It disappears.

The coexistence of experience and thought within human consciousness was accepted until the second half of the twentieth century, when behaviourists argued that, being unable to

---

15. Schopenhauer, A (1909) – *The World As Will and Idea*, Vol. 1, Kegan Paul, Trench Trübner & Co., p.26.
16. Schopenhauer, A (1909) – *The World As Will and Idea*, Vol. 1, Kegan Paul, Trench Trübner & Co., p.28.

conduct scientific studies on sentience, they had to sideline cognition.

An inveterate critic of European academia's *"cherished mediocrity"*, Arthur Schopenhauer was born in 1788, the son of a well-to-do German family; he attended Gottingen and Berlin Universities, where he studied medicine, philosophy, metaphysics, logic and psychology. His dissertation was *"On the Fourfold Root of the Principle of Sufficient Reason"*. He travelled widely, visited Italy, where he lived for one year, and spoke Italian fluently. He could also speak Spanish, Greek and several other languages.

Schopenhauer stresses the difference between *"ideas of perception and abstract ideas"*; of course, he is discussing sentience and cognition. He understands that that difference is among the most important facts regarding human consciousness.

"Besides the ideas we have as yet considered, which, according to their construction, could be referred to time, space, and matter, if we consider them with reference to the object, or to pure sensibility and understanding (i.e., knowledge and causality), if we consider them with reference to the subject, another faculty of knowledge has appeared in man alone of all earthly creatures, an entirely new consciousness, which, with very appropriate and significant exactness is called 'reflection'. For it is in fact derived from the knowledge of perception, and is a reflected appearance of it. But it has assumed a nature fundamentally different."[17].

---

17. Schopenhauer, A (1909) – *The World As Will and Idea*, Vol. 1, Kegan Paul, Trench Trübner & Co., p.66.

Here, the philosopher clearly states that *H. sapiens* is the only species which has *"an entirely new consciousness... called reflection."*. Complex cognition has not developed in any other animal species.

He saw with incredible clarity the causality between language and cognition: *"Speech is the first production, and also the necessary organ of his reason"*.

The connection with Eastern traditional thought is everywhere in Schopenhauer's work. It is totally pervasive, a constant; together with Gautama Buddha, he identifies cognition—which he calls *'willing'*—as the source of all human stress and sorrow:

"All 'willing' arises from want, therefore from deficiency, and therefore from suffering. The satisfaction of a wish ends with it; yet for one wish that is satisfied there remain at least ten which are denied. The satisfaction of a wish ends it; yet for one wish that is satisfied there remain at least ten which are denied. Further, the desire lasts long, the demands are infinite; the satisfaction is short and scantily measured out. But even the final satisfaction is itself only apparent; every satisfied wish at once makes room for a new one; both are illusions; the one is known to be so, the other not yet. No attained object of desire can give lasting satisfaction, but merely a fleeting gratification; it is like the alms thrown to the beggar, that keeps him alive today that his misery may be prolonged till the morrow"[18].

Schopenhauer understood, with Buddhism, that pleasure is never satisfied. Human beings tend to confuse pleasure with happiness. That leads to exactly the opposite: suffering.

Erwin Schrödinger

---

18. Schopenhauer, A (1909) – *The World As Will and Idea*, Vol. 1, Kegan Paul, Trench Trübner & Co., p.260.

. . .

"Reflection is the necessary copy or repetition of the originally presented world of perception, but it is a special kind of copy in an entirely different material.".
The World as Will and Idea -
Arthur Schopenhauer

Many books could be written about Schrödinger's contribution to knowledge. And—surely—many have. One of his finest achievements in physics was probably the wave equation—the behaviour of particles at the quantum level. But Schrödinger, a Nobel-prize-winning physicist, also ventured into biology, a science in which he collaborated in an important way with his discoveries regarding DNA, among other things. Until his death he continued to champion greater collaboration between physics, biology, and chemistry, especially to explain the emergence of life from inanimate matter. Some maintain that science has already solved some of those issues. In studying the origin of life, however, he proposed that living matter was ruled by aperiodic crystals, that is, it had a non-repetitive molecular structure. His were the first descriptions of DNA. He was, like Wittgenstein, a polymath.

Schrödinger became close friends with Einstein and, like him, was trying to find a unified field theory. They exchanged copious correspondence on the subject.

In 1935, Schrödinger introduced the famous metaphor of the cat in the box.

The friendship with Einstein suffered when Schrödinger had the idea that a rotating mass would generate a magnetic field, which he published without consulting Einstein. Einstein told him that it didn't differ much from his theory. After that they stopped writing for three years.

Schrödinger thought a lot about his idea that a spinning mass produced a magnetic field.

As mentioned above, in 1926 he elaborated the mathematical formula of the wave function—the way in which the position of a wave can be described as a range of positions. On that basis, Heisenberg proposed the Uncertainty Principle regarding the position of particles, and then, Neils Bohr presented an idea that combined much of what physics had seen up to that point: the position of a particle can be described as a wave, and the wave is actually the probability of a position. By marrying that idea with the Uncertainty Principle, Bohr concluded that the properties of particles are totally random. Uncertainty is fundamental in the universe. We have already seen that Einstein opposed this idea, saying that *"God does not play dice with the universe."* Niels Bohr replied, *"Don't tell God what to do."* It may have nothing to do with God, but what has been proven so far is that Bohr was right about nature.

Particularly drawn to the most complex problems involving mind and matter, Schrödinger gave a series of lectures on consciousness at Trinity College in 1956. They were published under the title *"What is life?"*. This is what interests us the most about his work: he explored the nature of self-awareness and subjective experience; he delved into biological processes and their philosophical implications. And he provided some important answers.

Life appeared on Earth roughly 4.3 billion years ago, a huge number. The planet is teeming with life. Before trying to understand consciousness, human beings, or mammals, we have to understand what life is.

Schrödinger describes how life does not follow the Second Law of Thermodynamics. Life does not decay towards equilibrium. He says:

"But, to reconcile the high durability of the hereditary substance with its minute size, we had to evade the tendency to disorder by 'inventing the molecule', in fact, an unusually large molecule which has to be a masterpiece of highly differentiated order, safeguarded by the conjuring rod of quantum theory. The laws of chance are not invalidated by this 'invention', but their outcome is modified. The physicist is familiar with the fact that the classical laws of physics are modified by quantum theory, especially at low temperature. There are many instances of this. Life seems to be orderly and lawful behaviour of matter, not based exclusively on its tendency to go over from order to disorder, but based partly on existing order that is kept up."[19].

The one amazing thing he says about life—which gives you a clear idea that his mind is that of a physicist—is that, according to the laws of physics, all living beings should die straight after birth, as all matter tends towards entropy. How do we stay alive? Schrödinger says: We eat negative entropy. I thought that was brilliant. A different way of thinking, which I biologist would never have considered. I know there is some multidisciplinary research but, I believe, it has generally been neglected.

Life is different from the rest of matter. Organic chemistry and

---

19. Schrödinger, E (1944)– *What is Life?*, Cambridge University Press, p.68

biology follow their own rules. The vital components of living creatures are completely different from the rest of nature. Life is based on internal order. The essence of inanimate matter is the exact opposite: total chaos.

So, if life does not follow the same rules as the rest of nature, which is inanimate, and life—as far as we know—is exclusive to our planet, then life, apart from being fundamental (I repeat, something we cannot explain), could be also be a random phenomenon (which would confirm Neils Bohr hypothesis).

The most important considerations on consciousness commence when he discusses the main principles of Western philosophy and science, what he calls *"the principle of the understandability of nature, and the principle of objectivation."*. Nature can be understood, yes, but that has to be done by hypothesising that there is a real world around us, he says.

In order to understand that world, according to Western thought—he says— I have to exclude the subject from the picture. I become an onlooker. I do not belong in that world. Once I accept that, I encounter several problems, not the least of them, that my body (and the bodies of other people—other spheres of consciousness) are part of the real world. So, the most important consideration here is that my own sentience is part of the material world.

Here, without saying it in so many words, Schrödinger prefigures Wittgenstein (the ineffability of some things) and our hypothesis of a 'hybrid consciousness'; he also puts himself squarely on the side of Eastern philosophy:

"Hence I am inclined to take them as something objective, as forming part of the real world around me. Moreover, since there is

no distinction between myself and others, but on the contrary full symmetry for all intents and purposes, I conclude that I myself also form part of this real material world around me. I so to speak put my own sentient self (which had constructed this world as a mental product) back into it – with the pandemonium of disastrous logical consequences that flow from the aforesaid chain of faulty conclusions. We shall point them out one by one; for the moment let me just mention the two most blatant antinomies due to the awareness of the fact that a moderately satisfying picture of the world has only been reached at the high price of taking ourselves out of the picture ... finding our world picture 'colourless, cold, mute'. Colour and sound, hot and cold are our immediate sensations; small wonder that they are lacking in a world model from which we have removed our own mental person."[20].

So, we can only 'understand' the world but not our sensations, because they cannot be 'understood'. Cognition understands only cognitive phenomena.

Schrödinger goes on to say that he is repulsed by the idea that:

"'... the world of science' has become so horribly objective as to leave no room for the mind and its immediate sensations."[21].

Further, he states:

"Mind has erected the objective outside world of the natural philosopher out of its own stuff. Mind could not cope with this gigantic task otherwise than by the simplifying device of excluding itself — withdrawing from its conceptual creation. Hence the latter does not contain its creator."[22].

I take it that when Schrödinger says that mind erects the

---

20. Schrödinger, E (1944)– *What is Life?*, Cambridge University Press, pp.118-119.
21. Schrödinger, E (1944)– *What is Life?*, Cambridge University Press, p.120.
22. Schrödinger, E (1944)– *What is Life?*, Cambridge University Press, p.121.

outside world of the philosopher out of its own stuff, he means that cognition only understands cognitive phenomena. Of course, he is quite aware that this is the construct of Western philosophy and science.

Schrödinger rejects the idea of the 'homunculus'. We know our consciousness does not reside within our body, that little man that observes the world from between our eyes. The location of our mind is only symbolic, he says. There is, however, a ceaseless movement of neurones and electrochemical impulses, thousands and thousands of contacts every split second within our nervous system. So that, in order for us to understand, there are things that move within us. But then, as a quantum physicist, he remembers that the boundary between subject and object is really inexistent:

"It is the same elements that go to compose my mind and the world. This situation is the same for every mind and its world, in spite of the unfathomable abundance of 'cross-references' between them. The world is given to me only once, not one existing and one perceived. Subject and object are only one. The barrier between them cannot be said to have broken down as a result of recent experience in the physical sciences, for this barrier does not exist."[23].

Quantum physics and Eastern mysticism appear to agree on this.

The next lecture is 'The Arithmetical Paradox: The Oneness of Mind'. Here Schrödinger commences by saying:

"The reason why our sentient, percipient <u>and thinking</u>* ego is met nowhere within our scientific world picture can easily be indicated in seven words: because it is itself that world picture."[24].

---

23. Schrödinger, E (1944)– *What is Life?*, Cambridge University Press, p.127.
24. Schrödinger, E (1944)– *What is Life?*, Cambridge University Press, p.128. * My underlining.

The thinking layer of consciousness *is* the one that analyses the world. Sentience and perception do not analyse. They are tools of that analysis. They are used in order to provide evidence of fundamentals. They are the ones for which there is no explanation. The conscious mind—he is saying—cannot analyse consciousness. In any case, Schrödinger speaks here of the paradox of many egos and one world. The traditional Western perspective: individual minds are separate and self-contained.

Schrödinger then introduces the Upanishads and Eastern mysticism into the picture, and suggests the alternative: minds and consciousnesses are unified. Minds are only iterations of the one consciousness. He is right, of course, in terms of the oneness of mind, but he errs in not excluding cognition. Natural consciousnesses are one. Cognition is artificial. As a human creation derived from language it does not share in their nature.

The other point he discusses in this lecture is the causality of consciousness:

"I find it utterly impossible to form an idea about either how, for example, my own conscious mind (that I feel to be 'one') should have originated by the integration of the consciousnesses of the cells (or some of them) that form my body, or how it should at every moment of my life be, as it were, their resultant."[25].

This is the phenomenon I discuss when I mention Dennett and his comparison of Gaudí's Sagrada Familia and the Australian termite mound. When one discusses consciousness, thinking in terms of neurones is counterproductive. It is short-sighted and does not take you anywhere because consciousness is not a purely biological phenomenon.

---

25. Schrödinger, E (1944)– *What is Life?*, Cambridge University Press, p.131.

Schrödinger's intuition seems logically correct to me. Consciousness appears to be a holistic phenomenon. Also, from an evolutionary perspective, organs tend to follow behaviour, not the other way round. *"The single nerve-cell is never a miniature brain"*, he says. I would venture that isolated lab-grown brain organoids (which didn't exist in Schrödinger's time) will never acquire full consciousness. Sub-minds are monstrous, as are plural minds. The first ones are artificial, the second ones have never existed. There are separations of the sensorium into different areas, that's all. That is Sherrington's paradox. The nervous system operates on the basis of the integration of many sub-systems.

Schrödinger submits two paradoxes of consciousness, one internal and one external. He then proceeds to indicate how to reach a solution:

"I submit that both paradoxes will be solved (I do not pretend to solve them here and now) by assimilating into our Western build of science the Eastern doctrine of identity."[26].

That is exactly what I am attempting to do here. The new element I introduce is the duality sentience/cognition.

Both, Western and Eastern schools of thought are partially correct because human consciousness has two layers. The cognitive layer asks questions and receives some answers (through religion, philosophy and science). As we mentioned before, some may be true, some not so true (I do not judge). As for the sentient layer, it asks no questions and receives no answers. It has no need because it is one with nature.

Schrödinger continues:

"The [cognitive] model is colourless and soundless and impalpable.

---

26. Schrödinger, E (1944)– *What is Life?*, Cambridge University Press, pp.134-135.

In the same way and for the same reason the world of science lacks, or is deprived of, everything that has a meaning only in relation to the consciously contemplating, perceiving and feeling subject. I mean in the first place the ethic and aesthetic values, any values of any kind, everything related to the meaning and scope of the whole display. All this is not only absent but it cannot, from the purely scientific point of view, be inserted organically. If one tries to put it in or on, as a child puts colour on his uncoloured painting copies, it will not fit. For anything that is made to enter this world model willy-nilly takes the form of scientific assertion of facts; and as such it becomes wrong."[27].

Briefly touching upon ethical considerations, Schrödinger tells us that life is valuable, but that nature does not value it. Nature does not issue ethical judgments.

"There is nothing good or bad but thinking makes it so. No natural happening is in itself either good or bad, nor is it in itself either beautiful or ugly. The values are missing, and quite particularly meaning and end are missing. Nature does not act by purposes."[28].

He is telling us that we are the witnesses and, he adds: *"The show that is going on obviously acquires a meaning only with regard to the mind that contemplates it."*. We believe that meaning and humanity share the same origin. Without the human mind there is no meaning.

The last of Schrödinger's lectures, *The Mystery of Sensual Qualities* is perhaps the most revealing. He appears to concur with the notion of a clear separation of the components of human consciousness:

"… the strange fact that on the one hand all our knowledge about the world, both that gained in everyday life and that revealed by the most carefully planned and painstaking laboratory experiments,

---

27. Schrödinger, E (1944)– *What is Life?*, Cambridge University Press, p.137.
28. Schrödinger, E (1944)– *What is Life?*, Cambridge University Press, p.138.

rests entirely on immediate sense perception, while on the other hand this knowledge fails to reveal the relations of the sense perceptions to the outside world, so that in the picture or model we form of the outside world, guided by our scientific discoveries, all sensual qualities are absent."[29].

Schrödinger explains in all kinds of scientific detail how the senses operate: the physicist's idea of yellow light is that it consists of *"transversal electro-magnetic waves of wave-length in the neighbourhood of 590 millimicrons."*. Notwithstanding that, *"The sensation of colour cannot be accounted for by the physicist's objective picture of light-waves."*.

The bottom line is that the explanation the physicist gives us cannot provide the sensation. Could a physiologist do better? Schrödinger doesn't think so. And the same thing that happens with yellow happens with sweet taste, musical notes, touch, hot and cold, smell, taste, etc. Nothing works. The paradox is that

"So we come back to this strange state of affairs. While the direct sensual perception of the phenomenon tells us nothing as to its objective physical nature (or what we usually call so) and has to be discarded from the outset as a source of information, yet the theoretical picture we obtain eventually rests entirely on a complicated array of various informations, all obtained by direct sensual perception.It resides upon them, it is pieced together from them, yet it cannot really be said to contain them."[30].

Let's say energy and matter are "fundamental" notions. Fundamental notions are notions we cannot explain, basically because we cannot fully explain how the universe started.

---

29. Schrödinger, E (1944)– *What is Life?*, Cambridge University Press, p.153.
30. Schrödinger, E (1944)– *What is Life?*, Cambridge University Press, p.163.

Energy and matter came with the original package. How? We don't know. Above, I propose that life is also a fundamental notion, albeit the result of the random behaviour of particles.

Schrödinger postulated that consciousness was also a fundamental phenomenon. I believe by that he meant just sentience. The principle is an important one in Eastern schools of thought: the interconnectedness of our (deceivingly plural) consciousnesses and the underlying oneness of nature. Schrödinger, however, believed that a fundamental aspect of the universe was a web of interconnected energy 'and information'. Eastern thought proposes the oneness of sentience but excludes cognition (i.e., information) from it. Again, I submit that cognition and information are exclusively human phenomena, and artificial to boot. Cognition and information cannot be included as part of any fundamental phenomenon.

I would guess that sentience (which is an exclusive quality of life) is as fundamental as life. As such, it is part of the oneness of nature. It is also within us, who are living beings. We are part of it. Our senses do not need communication or explanation because they are part of the oneness. They (and we) *are* the oneness. Sentience does not need time nor does it need quantities. It has no numbers. It is one.

Cognition, on the other hand, a by-product of meaning, which became language (an 'invention' according to Everett), was retrofitted into our consciousness. It is only subjective. It needs to understand, it needs to explain, it needs to communicate. Cognition comes with ethics and morals and beauty and judgement. Cognition comes with many other qualities that make us unique, and it is also the inquisitive component of human consciousness (I was going to write 'human nature', but—if I am correct—cognition is not natural). It has allowed our species to grow exponentially, it has allowed us to witness the universe, and to understand many natural phenomena.

The origins of the cognitive phenomenon, as explained above, came with language. After countless generations of combining sounds and adding meaning, a hominin uttered a combination of sounds with complex meaning (syntax). The interlocutor understood the meaning. Bingo. Humanity.

Let's see what Schrödinger said about time. One of the most important contributions of science—according to the physicist —was the *"gradual idealisation of time"*. Of course, that had happened long before science became science as a separate discipline. He refers to how humans passed from measuring cycles, to time-keeping, to an ideal concept of time. How time became part of human knowledge (or, to put it in other words, how we invented time). Schrödinger mentions Plato as the first to contemplate the possibility of a *"timeless existence"*. He then discusses causation:

"Time is the notion of 'before and after'. ... The notion of 'before and after' resides on the 'cause and effect' relation. We know, or at least we have formed the idea, that one event A can cause, or at least modify, another event B, so that if A were not, then B were not, at least not in this modified form."[31].

Schopenhauer uses different words to say the same thing:

"This simplest form of the principle [of sufficient reason] we have found to be time. In it each instant is, only in so far as it has effaced the preceding one, its generator, to be itself in turn as quickly effaced. The past and the future ... are empty as a dream, and the

---

31. Schrödinger, E (1944)– *What is Life?*, Cambridge University Press, p.147.

present is only the indivisible and unenduring boundary between them."[32].

The effect cannot precede the cause. That is fine, but then Schrödinger admits that causation—which is clear in mathematical language—becomes muddled because:

"It then follows that the above-mentioned discrimination between 'before and after' or 'earlier and later' (based on the cause-and-effect relation) is not universally applicable, it breaks down in some cases. This is not as easily explained in non-mathematical language. Not that the mathematical scheme is so complicated. But everyday language is prejudicial in that it is so thoroughly imbued with the notion of time — you cannot use a verb (verbum, 'the' word, Germ. Zeitwort) without using it in one or the other tense."[33].

The fact is that there are events that are neither earlier nor later than A.

"So if we want to make this relation, and not a linguistic prejudice, the basis of the 'before and after', then the B and B' form one class of events that are neither earlier nor later than A. The region of spacetime occupied by this class is called the region of 'potential simultaneity'."[34].

"Potential simultaneity" becomes the liberation from causation.

Within sentience, time exists solely as change, but then human cognition explains it through causation and then measures it, and then it really becomes time. Time-keeping is not an explanation of change, it only measures it.

Schrödinger then discusses the unidirectionality of time. Time goes from past to future, not the other way around. That is

---

32. Schopenhauer, A (1909) – *The World As Will and Idea,* Vol. 1, Kegan Paul, Trench Trübner & Co., p.30.
33. Schrödinger, E (1944)– *What is Life?,* Cambridge University Press, p.148.
34. Schrödinger, E (1944)– *What is Life?,* Cambridge University Press, p.149.

linked to the Second Law of Thermodynamics: the entropy of isolated systems left to spontaneous evolution cannot decrease, as isolated systems tend to thermodynamic equilibrium, i.e., they rest.

Without interference, whatever is hot always becomes colder, not the other way around (mechanical or statistical theory of heat).

According to Schrödinger, the statistical theory of time:

"... has an even stronger bearing on the philosophy of time than the theory of relativity. The latter, however revolutionary, leaves untouched the unidirectional flow of time."[35].

In any case, he concludes that—because time is a creation of cognition—physical theory suggests that mind cannot be destroyed by time. That goes without saying. Time is a human construct. It's the other way around. Without consciousness, time will cease to exist. It did not exist before consciousness.

---

35. Schrödinger, E (1944)– *What is Life?*, Cambridge University Press, p.152.

A CONTEMPORARY PERSPECTIVE

ADAM FRANK, MARCELO GLEISER & EVAN THOMSON

*T*hus far, I have emphasised the work of nineteenth and twentieth-century thinkers. But there are three contemporary researchers—one physicist and two philosophers—who bluntly but elegantly question the current physicalist scientific worldview. I know there are other researchers rebelling against convention but these cannot be ignored: their comprehensive approach is something rare.

In their brilliant book *"The Blind Spot"* (2024), Frank, Gleiser and Thomson bring Eastern thought and quantum physics again to the fore, and suggest alternatives.

The introduction to their book points to some ideas that first appeared during the Enlightenment:

"According to that worldview, nature is nothing but shifting spatiotemporal arrangements of fundamental physical entities. In this perspective, the mind is either a derivative physical assemblage or <u>something radically different from nature altogether</u>.[*,1]*"*

Throughout the book, again and again, the authors go directly

---

1. Frank, A *et al* – *The Blind Spot* (2024), MIT Press, p.vii. * My underlining.

to the point: the scientific world needs to change its approach to the mystery of consciousness and in order to achieve that it must forsake materialism and include Eastern thought, dualism and quantum physics:

"Each of the cases just mentioned—cosmology and the origin of the universe, quantum physics and the nature of matter, biology and the nature of life, cognitive neuroscience and the nature of consciousness—represents more than an individual scientific field. Collectively they represent our culture's grand scientific narratives about the origin and structure of the universe and the nature of life and the mind. They underpin the ongoing project of a global scientific civilization. ... In short, although we have created the most powerful and successful form of objective knowledge of all time, we lack a comparable understanding of ourselves as knowers. We have the best maps we've ever made, but we've forgotten to take account of the map makers. Unless we change how we navigate, we're bound to head deeper into peril and confusion."[2].

At the beginning of this point I say that the authors 'bluntly but elegantly' question current scientific trends. Still in the Introduction to their book, they make quite clear what they mean:

"In plumbing the depths of experience, we have no intention of downplaying the success and value of science. We reject science denial, but we also reject scientific triumphalism. Our quarrel is with a particular, misguided conception of science, one that has come to be built into our present scientific worldview but isn't an essential part of science. This misguided conception, which we delineate in Chapter 1, is essentially a philosophy of science based on certain metaphysical assumptions about nature and human knowledge. We argue that science doesn't require this philosophy and that given its failures, we should jettison it and move on."[3].

---

2. Frank, A *et al* – *The Blind Spot* (2024), MIT Press, p.ix.

3. Frank, A *et al* – *The Blind Spot* (2024), MIT Press, p.xvi.

Unfortunately, there were schools of thought during the twentieth century, for which materialist science had become the end-all solution to knowledge. Fighting against religion, scientism was transformed into religious dogma. You either believed, or you were the object of ridicule. Hitchens, Dawkins, Dennet and Harris were, to some extent, responsible for something that became the norm. Nobody wants to downplay the success of science. Science, however, has to remain flexible; it cannot become dogma. When scientists say that materialism is the only acceptable philosophy of science, then something is wrong.

I believe the main test of this situation is the study of consciousness. In order to understand consciousness, science needs to accept its limitations. If you can imagine that you are a being that lives in an aquatic medium—say a whale, or a dolphin—and you need to explain how humans behave, you would need to change your thought patterns, you would need to consider an alternative paradigm. You would not be able to compare any phenomenon that occurs under water to how terrestrial animals behave. The fact that gravity acts differently in media with different density—like air and water—would make your job extremely difficult, if not impossible. If you had to study how human beings run, for instance, hydrodynamics would not be of any assistance.

In the case of consciousness, science is not equipped to deal with it because science itself operates within a cognitive medium. The same thing happens with philosophy; as Wittgenstein very well stated: when you cannot explain something, the best thing you can do is to remain silent. Complex cognition is a unique, human, phenomenon. To make matters even more difficult, all indications are that sentience operates as a quantum system, to which the taxonomy observer/objective reality does not apply. Humanity appears to be a superorganism that functions within an interconnected reality. These are phenomena that the current scientific orthodoxy does not appear prepared to accept.

Frank, Gleiser and Thomson place the origin of the gap between science and consciousness— what they call the 'Blind Spot'—in pre-Enlightenment times. Is it a coincidence that the seventeenth century was when Rene Descartes introduced substance dualism?

"The problem goes back to the rise of modern science in the seventeenth century, particularly to the bifurcation of nature, the division of nature into external, physical reality, conceived as mathematizable structure and dynamics, and subjective appearances, conceived as phenomenal qualities lodged inside the mind. The early modern version of the bifurcation was the division between 'primary qualities' (size, shape, solidity, motion, and number), which were thought to belong to material entities in themselves, and 'secondary qualities' (color, taste, smell, sound, and hot and cold), which were thought to exist only in the mind and to be caused by the primary qualities impinging on the sense organs and giving rise to mental impressions. This division immediately created an explanatory gap between the two kinds of properties." [4].

Not only do these three authors question current scientific views; they also question how we consider classical reality as 'the way things are', and the manner in which traditional Western philosophy rejects the holistic Eastern perspective of reality:

"An intrinsic property is traditionally understood as a property that something would have even if it were the only thing in the universe or the only thing in existence. Does that idea even make sense? Not if you think that something is what it is only by virtue of its belonging to a web of relations. Why not say that relations deter-

---

4. Frank, A *et al* – *The Blind Spot* (2024), MIT Press, p.193.

mine the occupants of the relations, after the fashion of relational quantum mechanics (see chapter 4)? Or that relations and occupants are mutually interdependent? Madhyamaka (Middle Way) Buddhist philosophers, over centuries of discussion with their Indian philosophical interlocutors, have given compelling reasons to reject the intelligibility and existence of intrinsic properties. Their arguments have inspired analytical philosophers and quantum physicists to maintain the primacy of relations over entities with intrinsic properties. In any case, the above argument turns on the metaphysical concept of an intrinsic property, a concept that is very difficult to make sense of."[5].

What the authors are discussing here is not just an abstract philosophical question. It directly applies to the way in which current science concentrates on the individual human brain, to the exclusion of any cognitive interactions it may require in order to function within culture. It directly applies to any intrinsic properties of the individual brain, which evidently is a tiny component of a global superorganism: humanity. Consciousness cannot—and does not— 'emerge' from neurones or synapses in the individual brain. There are two separate phenomena in human consciousness: one of the two components that originate the phenomena is cognition, the gel that makes human interconnectedness possible.

One last point about Frank, Gleiser and Thomson. I read the book as their attempt to change current scientific worldview. They do it by presenting extremely credible arguments. They emphasise inclusion as the only way we are going to understand human consciousness:

"We began this book by pointing to a paradox. Science tells us that we human beings are peripheral in the cosmic scheme of things, but also that we're central to the reality we uncover. We're a tiny

---

5. Frank, A *et al* – *The Blind Spot* (2024), MIT Press, p.214.

contingent presence in a vast universe, yet our experience is ubiquitous in scientific knowledge. We've seen many examples in the previous pages.

Instead of trying to avoid this paradox, we can and should embrace it. We're the authors of the scientific narrative, and we're characters within it. As authors, we create science. As characters in the narrative, we're a minuscule part of the immense cosmos. This is how we must portray ourselves based on what we've discovered in cosmology and biology. Still, we must not forget that these tiny characters—we ourselves—are also the authors of the narrative. We've written it using the tools we created as a community of learners in the collective scientific workshop. Such is the strange loop of scientific knowledge."[6].

This may sound anthropocentric to some. The fact remains that, as far as we know, human consciousness is the only witness in the whole cosmos. Inclusion appears to be the only way to understand ourselves and the universe.

---

6. Frank, A *et al* – *The Blind Spot* (2024), MIT Press, p.251.

ANOTHER PHILOSOPHY

EASTERN SCHOOLS OF THOUGHT

"How hard, then, and yet how easy it is to understand the truth of Zen! Hard because to understand it is not to understand it; easy because not to understand it is to understand it. A Master declares that even Buddha Sakyamuni and Bodhisatva Maitreya do not understand it, where simple-minded knaves do understand it."
D.T. Suzuki - *An Introduction to Zen Buddhism*

"The over-all number of minds is just one. I venture to call it indestructible since it has a peculiar timetable, namely mind is always now. There is really no before and after for mind. There is only a now that includes memories and expectations."
Erwin Schrödinger - *What is life?*

This chapter will deal with an Eastern view of consciousness, which is something completely different from the treatment of consciousness in Western philosophy.

For millennia, East and West have held opposed—seemingly irreconcilable—views of what human consciousness is. The

East appeared to have chosen wisdom, and the West, knowledge. On different occasions throughout history, the East kept on choosing sentience and meditation; the West chose cognition and inquisitiveness. Experience versus thought. Both profoundly valid.

By asserting that human consciousness has two separate natures, i.e., that it is a hybrid phenomenon, I am not actually saying that both, Eastern and Western schools of thought are correct. On the contrary, but they are both needed to understand our mind. Thus far, Western science and philosophy have tended to ignore the East altogether.

Zen is not a religion, nor is it a philosophy. It has no definition. Zen is a return to nature.

I will try to explain how Zen unravels the components of human consciousness. The process began with Buddha in Nepal and took centuries to evolve, until Boddhidharma distilled its essence in China. His pupil, Dōgen Zenji, picked up the baton, and this eventually led to the founding of the Sōtō school of Zen in Japan.

Zen practitioners go where science cannot go. Which means that, sometimes, not understanding is all right. You will be the judge as to how logical or credible the conjectures are.

Gautama had guessed the dual nature of human consciousness. Sentience, he discovered, was the key to *nirvana;* utter wisdom came only through direct sensorial experience and nature. Contentment has to do with accepting the present and enjoying it. Within sentience there is nothing to be explained. No future means nothing to stress about.

Buddha said that the highest state a human can aim for is ecstasy, which is achieved through meditative concentration.

Nirvana was the only form of salvation. Buddhism rejected the separate existences of original consciousness and matter, subject and object, soul and deity. There was one reality. Only a dream that has no dreamer. The dream, Buddha said, is surrounded by nothingness.

Eastern schools of thought maintain that everything is connected. There is a clear logic to that assertion. Much more than people normally understand. At a much more intimate level. We are matter. We are made of the same stuff as our surroundings. That is not science, but it has a scientific explanation. Tom Chi, an astrophysicist and polymath, explains it brilliantly. I recommend his videos. Maybe a Western mind needs an explanation like that. It needs to 'understand' why things are the way they are.

As with the rest of other animals, our hearts beat in order to carry a molecule called haemoglobin through our bloodstream. Every molecule of haemoglobin has a single iron atom (Fe^{II}). So iron is absolutely essential for us to be alive and stay alive. In the universe, iron is only created is through the formation of supernovas and other massive stars.

At the beginning, the universe had elements like hydrogen and helium; but it had no iron whatsoever, so life could not have existed. The collision and explosion of hundreds of thousands of stars and galaxies— which happens because of gravity—produces iron. That is the same iron that runs through our veins. The fact that our veins carry an element created in faraway galaxies is difficult to comprehend, but there is no other explanation for that vital element within us.

Our planet appeared over four billion years ago. It was a very different place. At that stage, there was no oxygen on Earth. The atmosphere had as much nitrogen as it has now, but no oxygen. There was a lot of carbon dioxide. Only single cell organisms could live here. Some two odd billion years ago, there was an organism, called cyanobacteria, which could produce photosynthesis. That means they would take energy from the sun and would transform carbon monoxide into oxygen. They would synthesise light. Over billions of years, those bacteria produced the oxygen we now have in the atmosphere and we can now breathe. The seas were produced first, and then the ozone layer. Without it there could be no multicellular life on Earth. And only after the Cambrian explosion could there be life on land. All those bacteria that lived so long ago were the origin of our lives. Cyanobacteria still exist in the plants we eat, which are sources of photosynthesis. We breathe in the same oxygen plants breathe out.

These few previous paragraphs attempt to provide a Western explanation for the beginning of life. It took a long time for sentience to develop within living beings. And a much, much longer time for cognition to become part of human consciousness.

~

Human beings are curious; scientists and philosophers are curious. That—many would say—is part of the cognitive component of our consciousness. Actually, purely sentient beings can be curious: cats are curious; many corvids, like crows and magpies, are curious; bears are curious. They are not inquisitive, though. What human beings do that differs from other animals is questioning. No other animal asks questions or gives explanations, or not that we are aware of. But it is easy to explain. No one can ask questions without language.

Questioning is one of the expressions of curiosity. It is also an expression of doubt, of perceived possibility, or probability. Our questions have good and not so good qualities. The good qualities are the ones involved in science and philosophy:

they are a part of our quest for knowledge. The bad qualities involve doubts about ourselves, fear or fears, anxiety, etc.

Questions in Zen, however, receive no answers (or no logical answers). Koans are famous for being unsolved puzzles. *"What is the sound of one hand clapping?" "Why did the Boddhidharma come from the West?"* That is, those questions are not well received. You can ask, but the answer will probably not be to your satisfaction. Zen Masters will tell you that you will not be able to find out about Zen through questioning, because sentience cannot be explained. It can only be demonstrated.

Zen doesn't explain. Anything. Zen accepts life as it is, without questioning or asking for explanations. Zen gives you an example of how things are: the flag flutters in the wind. Is it the flag that flutters or the wind that blows? Neither. It's only your mind trying to explain something. It is your mind that moves. The flag and the wind may move, but they need no explanation. They just are.

I can explain a theory, an equation, a formula. I can explain a situation, a puzzle or a problem.

I cannot explain to you what being wet feels like, that lies in the realm of sentience. I can say that your skin is waterproof. I can say that if you jump into the water, you will go inside the water, but the water will not dissolve you; it is possible that you may swallow some, but it will not go into you otherwise. None of that will give you the experience of feeling wet. Diving or walking in the rain will. You cannot understand the act without experiencing it.

I can try to explain C minor to you. I can say that its key signature consists of three flats. Nothing will do, except hearing it. The same thing happens with the colour magenta. It is a light mauvish-crimson which is one of the primary subtractive colours. It was named after a battle that took place in Italy during the Napoleonic Wars. Explanations will mean nothing to you until you see the colour with your own eyes.

There are phenomena that cannot be explained with language or demonstrated through formulas. I repeat, Ludwig Wittgenstein—arguably one of the greatest philosophers of the twentieth century—stated *"Whereof one cannot speak, thereof one must be silent."* (see the chapter on *Ludwig Wittgenstein*). Maybe he wasn't quite clear enough. Perhaps I could paraphrase or clarify the concept in the context of Zen: "Whereof one cannot speak, thereof one must allow sentience to take over.".

But let us go back to cultural differences between East and West. Theologically and philosophically, East and West had similar beginnings. Buddhism was an offshoot of Hinduism, and Christianity an obscure cult that developed from Judaism. The results—however—were starkly opposing, irreconcilable, views of ourselves, of nature and the universe, both of which are partly correct. They are correct because of the hybrid nature of our consciousness.

The Hebrew Bible defined humans as something totally separate, above from the rest of creation. God put Adam in charge of all animals. Christianity, influenced by Greek thought, gave humans an immortal soul. When good Christians died, they went to Heaven, with God. They were individual demigods. That individuality was further emphasised by Martin Luther, who accepted no intermediaries between Man and God.

Humans could study a reality that was separate from them. Reality was objective. Nature was objective.

Secretly, astrology and alchemy gave birth to astronomy and chemistry, and other disciplines followed. Science crept slowly out of a nebula of superstition. Newton and Descartes —an alchemist and a magician—became the fathers of science and philosophy.

Things developed strangely. It was like a dance in which religion, art, philosophy, and eventually science, acted as the four bases of Western DNA, which intertwined to form the double helix of the culture. The West had opted for cognition.

At the beginning, civilisation in the East—especially in the subcontinent—had grown out of all proportion: from small clans and trading settlements into major conglomerations of human beings. Cities like Mohenjo-Daro and Harapa had tens of thousands of inhabitants, which was hard to conceive at the time. The jump from hunter-gatherer groups to agricultural communities to cities had been fast and fraught with difficulties and problems.

There were castes and big differences in social stations; cities had the equivalent of gated communities, where the more affluent enjoyed their privileges. Many found that life in the city was stressful, or oppressive, or both. Competition and the need to acquire material things became too much and some either never integrated or eventually decided to drop out of the "rat race". There was a nostalgia for the old, more natural, life. Many reacted in a way very similar to the hippie movement of the sixties and seventies. Hippie interest in the East was no coincidence.

Fakirs, yogis and other mendicants appeared in the environs of big cities. Siddharta Gautama expressed the feelings of monks and other outcasts; he preached a reactive kind of non-attachment (to things or people). His "Middle Path" rejected ascetic life as much as carnal life and desires. Gautama had found a remedy against suffering. In large societies, cognition (and suffering with it) had been growing through the exponential use of thought, language and writing.

Gautama became the icon of the movement. He was declared the "Buddha" (the Awakened One).

Buddhism, born as a way of life, grew out of Nepal, meandered through India and then pushed into the continent where—from a philosophy—it became a religion with rites, scripture and all the other trappings of religion. By the sixth century CE, when the patriarch Boddhidharma travelled North, Buddhism reached China. There, it coexisted with Confucianism and Tao.

After nine years meditating before a wall, where his image became imprinted—or so the legend goes—Boddhidharma founded *Ch'an*, a meditation sect. It involved sitting cross-legged in what is known as the "lotus position". That position was called *"zazen"*, later shortened to Zen, in Japan.

How Zen entered Japan and became so influential in the country is a long story. Here, we can only say that, by the time it arrived, it had been distilled into the essence of Buddhism.

We know that, after countless millennia, the layers of human consciousness had become so enmeshed that it appeared impossible to peel one of them off. This is where Zen meditation holds its secret: meditation practice gradually dissolves cognition and exponentially heightens sentience; that is called *"satori"*. The fact that this is at all possible demonstrates that those strata, those layers, are discrete and have different natures.

We live stressing about the future and regretting things that we did or did not do in the past. When you focus your life in the present, you realise those problems do not exist.

As you practise Zen meditation and apply it to your daily life you will experience, at a certain point, something Masters call *"luminosity"*. That is the moment you realise that the activities of your body and your mind are not totally separate and you gain an incredible insight. Body and sentience are one and the same. During meditation, cognition and time slowly dissolve and, with them, your own identity disappears. You become wise; you are part of *'oneness'* again. Actually, you and *'oneness'* are the same. You and all other people (and nature) become one. Time does not exist. When that happens, no explanation is needed. Thought is gone.

Buddhism in India and China

"Of course you are uncertain, Kalamas. Of course you are in doubt. When there are reasons for doubt, uncertainty is born. So in this case, Kalamas, don't go by reports, by legends, by traditions, by scripture, by logical conjecture, by inference, by analogies, by agreement through pondering views, by probability, or by the thought, 'This contemplative is our teacher.' When you know for yourselves that, 'These qualities are bad; these qualities are blameworthy; these qualities are censured by the wise; these qualities, when adopted & carried out, lead to harm & to suffering' — then you should abandon them."
Gautama Buddha - *Kalama Sutta*

The fact that Buddhism appeared in India when there were already large cities in the subcontinent should come as no surprise. For a long time, human beings lived in small clans. Up to that point, human instincts had adapted without much problem to living in groups larger than clans but smaller than cities; tribes appear to have been a —still comfortable—limit for the individual. Before then, the hunter-gatherer setting was a situation in which humans functioned very well. Civilisation complicated matters a lot.

Like any other social animals, humans, when placed in crowded situations, suffer all kinds of issues, for instance, competitive stress derived from a lower social status or rank, depressive moods, or anxiety, and even an inefficient immune system. The desire for prestige or fame, wealth and material possessions occurs only in large civilisations. In smaller groups, it is non-existent.

It is quite possible that the transition between living in hunter-gatherer groups, —even agricultural groups— and life in a city may have been quite traumatic for some. Apparently, cities in India had the equivalent of gated communities, where the more affluent groups enjoyed their privileges, to the exclusion of the majority, which had to labour without much freedom.

The appearance of fakirs, yogis and other mendicants around the first Indian cities could not have been a coincidence. These were people who could not integrate, and people who found that the competition they had to suffer in order to possess material things within the city was not worth the effort. Life was suffering. There was kind of a nostalgia for a simpler type of life.

A mixture between hippiedom and socialism, Buddhism surged from Hinduism as a rejection to the brutality of civilisation. Here, the comparison with Christianity is also very interesting.

Gautama combined the feelings of many of those—voluntary or involuntary—outcasts. He agglutinated their emotions and provided the motives they needed for their choice. What he preached was a reactive kind of non-attachment (to things or people). A return to the more basic existence they craved, but with a motivation. Civilisation required a lot of thought (and a life in which sentience was dimmed); human beings could find a much more enjoyable life through basic experience and without material possessions. Nothing lasts forever, things change. Why attach ourselves to things that will not last? At what price? But the most important thing he preached was acceptance to whatever happens in the present. Acceptance was wisdom. Wisdom was the closest to happiness humans could hope for. Contentment.

We know that Buddhism in India—like Christianity in Judea—began, literally, in the periphery of the main religion, in this case, Hinduism.

While Christianity rejected sacrifice, revenge, and violence in general—and we can see that in Matthew's spurning of Talion Law and in Christ's advice to turn the other cheek, among other things— it generally accepted the growth of society and civilisation in Judea, and asked people to join the sect.

Although he lived in a cave in the desert, John the Baptist, Jesus' relative and mentor, invited people into the community and, in accepting them, gave them a communal identity. He baptised crowds in the Jordan river.

Gautama Buddha, on the other hand, became a hermit himself and rejected larger societies altogether. What is clear from his teachings is that he did not believe that thought, logic, conjecture, inference, or analogy could lead anywhere near peace and acceptance of reality. Nothing but direct experience provided a valid answer to the needs of the individual. The communion he was looking for was with nature rather than with society. He rejected anything he considered vaguely artificial.

It is easy to conclude that Gautama had guessed the dual nature of human consciousness which, in larger societies, was reaffirmed by the introduction of extensive use of thought, language, and writing. His rejection of anything related to cognition and his quest for peace on the basis of a return to direct sentience appears to confirm that. But his greatest discovery was the inner being and its oneness with nature. Sentience— he discovered—was the way to *nirvana*. Utter wisdom comes only from direct experience.

Of course, the different philosophies adopted by Gautama and Jesus produced very different results. The West ended up with a highly individualistic society whereas, in the East, the individual accepts the rules of the group to a much larger extent. Intersubjectiveness is of paramount importance.

The paradox is that Christianity insists on a 'communion of sorts'... the collective is always placed above the individual, whereas Buddhism strongly advises and looks for an inner spiritual experience apparently centred on the individual.

Above, I quote Gautama, when he tells the Kalamas that, if they doubt, they should abandon any practices that are not what they had expected. The Kalamas had been visited by Brahmins, who said that their truth—Brahmin truth—was the only real one.

Gautama told the Kalamas that, if they doubted, if they thought what the Brahmins had told them was not true, if they thought it could lead to suffering, they should abandon that belief. He told them to go and find out for themselves. Gautama also advised them against *"scripture, ... logical conclusions... or ... thought"*. By that he meant, *"do not accept explanations, only experience."*.

Buddhist scripture sees Gautama as someone who really emerges from Hinduism with a different search altogether. He is like the Martin Luther of the East. The revelation comes directly from the gods. At one point, in the old Pali texts, Brahma descends from the heavens with a large retinue and asks Gautama to commence preaching, as his preaching would save mankind. Gautama agrees.

But Gautama—we find out later—is not looking for God. That requires imagining that there is a God or gods. Gautama preaches focusing on something that is as far away from the divinity as possible: he suggests looking at oneself and searching for the inner being. The world—he says—is contingent. Change is inevitable, as is suffering. There is inevitability in being born, growing up, getting sick, growing old and dying. All of that has to do with time. None of that has anything to do with the present.

After agreeing with Brahma's request, he goes to Vanarasi, begs for alms and looks for five monks who were ascetic friends of his. He shows them the Middle Path, which lies between carnal life and ascetic life. The monks convert to Buddhism.

In the Middle Path, which he preaches, Gautama says there are four truths: suffering, the cause of suffering, the end of suffering, and the path that leads to the end of suffering.

One interesting principle that emerges from the Middle Path is that it rejects asceticism—in which the individual uses prayer or mortification, and the body is made to suffer to achieve a higher end— and it also rejects carnal desires, which eventually result in suffering. What is left is a natural life of acceptance with no indulgence but no denial either. The Middle Path looks for a sophisticated kind of reunion with nature: the discovery of the inner being.

∼

There must have been a real Buddha as well as the mythical one. In India, chronology was not important at the time (they still appear not to be that keen on lists of kings and battles). So, all we have left is the mythical Buddha. In any case, it would be a waste to try to narrate what the legend says about him. For instance, telling a Western audience that the Buddha entered his mother from one of her sides in the shape of a young white elephant with six tusks would not mean anything to them. The mother, of course, was a virgin.

We know some details that appear to be real, like the fact that Siddharta Gautama was twenty-nine years old when he commenced preaching. We also know that, before that, he had several teachers. Contrary to what legend says, his father appears to have been a rich Nepalese landowner from Kapilavastu, not a king.

What preceded Buddhism was a Hindu philosophy called *Samkhya*. Basically, *Samkhya* is a dualist conception of human reality: there is *Purusha*, a male witness-consciousness, which cannot be defined or analysed, and *Prakriti*, a female consciousness, which includes cognitive, moral, psychological, emotional, sensorial and physical aspects of reality. Between *Purusha* and *Prakriti* lies *Buddhi*. Is that the germ of the Middle Path?

In *Samkhya* we can see a germ of Buddhism. But there are also big differences. *Samkhya* delights in analysing and enumerating (I am trying not to include any of the lists in Pali or Sanskrit here as they only cause confusion).

Purusha sounds a bit like *nirvana* or *satori*. Also, there appears to be a linguistic connection that is interesting to note: *Prakriti*, the word, has a strong similarity to the *Prakrits*, the languages that descend from Sanskrit. So, for *Samkhya*, language (mostly cognition?) and real wisdom seem to be on opposite sides of reality.

Also preceding actual Buddhism is *Vedānta*, a philosophy of wisdom and salvation that for millennia was transmitted orally. It is a kind of pantheistic monism. Only the present is important.

The *Vedānta* claims that man continues living exactly the same after *nirvana*, like the potter's wheel keeps on spinning after the pot is finished. One famous Zen saying is *"Before satori, chopping wood and carrying water; after satori, chopping wood and carrying water"*. Enlightenment does not change the enlightened. In many ways, it's imperceptible.

But there is no need to dwell much longer on these ancient practices. For our purposes, the Indian religions, philosophies and disciplines that preceded Buddhism only complicate matters.

Even the followers of Buddha made things more difficult by adding rites, enumerations, roles, explanations and superstitions. What interests us is how Zen emerged from all these convoluted theories. How Zen acquired the elegant simplicity it has.

The essence was there: Buddha himself rejected abstract debates and had come up with the famous parable of the man who, wounded by an arrow, does not want to have it removed until he knows the name, caste, nationality and other details of the archer who fired the shot. Buddha said that doing that was running the risk of death. "I am here to teach how to remove the arrow" he said. According to him, all other speculations were useless.

Buddha claimed that the highest state a human can aim at is ecstasy, which is achieved by means of meditative concentration. *Nirvana* is the only form of salvation.

Gautama made a clear distinction between relying on, or attaching ourselves to, something or someone outside of our true self. It may bring temporary pleasure but that is all. He preached non-attachment. Pleasure vanishes because it is artificial; paradoxically, it is based on separation. It comes from the outside in. Joy flows from the inside out. Attachment brings transient pleasure, which will eventually disappear and bring pain. Non-attachment means lasting joy.

Identities need to be satisfied, they need external possessions, that is, things, and people, and riches. But there are other possessions that are intangible, like social status and physical appearance, fame and recognition. In any case, once the individual needs possessions, the need will never disappear, that individual will need more and more. When you take those away—and that is bound to happen because possessions and relationships are transient—the individual suffers.

Gautama never intended to found a religion. His sole aim was teaching how to free his disciples from the world of appearances. *Nirvana* breaks the cycle of reincarnations. That is the end of suffering.

Advaida Vedānta is considered as a school which is still part of Hinduism. The difference with *Theravāda* Buddhism is that *Vedānta* claims there is no difference between subject and object (the unity of *Atman* and *Brahman*), whereas *Theravāda* goes one step further and claims the non duality of reality. Buddha rejected the *Vedas* (or texts). However, his teachings resulted in texts like the *Pali* Canon, or *Tripitaka* (three *Pitakas* or chapters), which is a massive (sixty-odd volumes) collection of teachings and doctrine.

Hinayana doctrine (the Little Vehicle) is the first version of Buddhism proper, *Mahayana* (the Great Vehicle) makes its appearance around the second century CE. *Mahayana*, pretty much like *Samkhya,* is full of divisions and subdivisions, enumerations, negations and affirmations. Both doctrines share the notions of impermanence of the self, suffering and the unreality of the ego. They also share the Four Noble Truths, Karma and the Middle Path. Both doctrines reject causality. Things just happen. The individual does not exist in isolation.

In Tibet, *Mahayana* becomes Lamaism, which is institutionalised as a theocracy, where there are many rites and hierarchies. The Dalai Lama is the head of state and the Pantchen Lama is the head of government. The current Dalai Lama lives in exile in India, after the occupation of Tibet by communist China in 1949.

How Buddhism travelled eastwards, towards China, becomes an interesting step in the history of the philosophy, which in certain places becomes a religion. What happens in China is fairly complicated. The approximate date when that happened is not quite certain. In any case Buddhism had to face a centuries-old tradition of Confucianism. Taoism was another important force founded, like Confucianism, in the sixth century BCE.

What is certain is that Buddhism became really influential around the sixth century of our era. Bodhidharma arrived in China in 526 CE. There are many stories about the patriarch in China. According to legend, he spent nine years meditating before a wall, where his image became imprinted; he later founded *Ch'an,* the meditation sect. Meditation involved sitting cross-legged, in what is called the "lotus position", what became *zazen* in Japanese, and engaging in concentration using several different methods, like repeating a *koan* or a mantra, or silently reciting the name of a Buddha.

Ch'an simplified. There was a total rejection of Indian thought. The convoluted explanations and the sixty-odd volumes of the Pali Canon, and many other ancient texts were left behind. The path could not be explained with words. Even the statues of Buddha were not there to be worshipped. *Ch'an* meant that the person had to experience enlightenment directly. There was no substitute for experience because it was all that was.

Ch'an, as we know, went to Japan, where it became Zen.

So, there being no more words, no more Pali Canon, no more deities to be worshipped, what was the result of that transformation? You cannot call Zen a religion, as there is no worship involved. You cannot call Zen a philosophy either, because there is no rational quest to explain anything. Buddhism grew out of a philosophy that rejected Hinduism, a polytheistic religion—and it actually became a religion itself, again, in Tibet. Then it crossed into China, where it became a way of life.

Zen is, in many ways, an oxymoron. Far from being a religion, its practitioners are mendicant (but working) monks of some sort, if that makes any sense. It absolutely rejects thought, or any form of logic or cognition. Without allowing questions, it provides the most transcendental answer to any human individual: how to enjoy life and find a meaning for it. Further than that, although it is a type of mysticism, it establishes that you should not reject your body, on the contrary, it tells you that you should look for your inner body.

To a mind used to Christian mystics, individuals like St Catherine of Siena, St Francis of Assisi, St Rose of Lima, or St Ignatius of Loyola, people who aspired to a spiritual union with God and rejected their flesh as the core of sin, the emphasis on physicality that Zen advocates is difficult to understand. Looking for the union between your body and nature—reality and being—doesn't quite seem like mysticism. The difference, perhaps, lies in the fact that Zen does not recognise a spiritual or supernatural reality. What replaces mortification in Zen is discipline and method. Both, together, eventually produce a strong but different type of moral character in the follower.

Christian mystics wish to achieve a trance and that trance takes them to a special union with God, and that follows a certain logic. There is a method: denial of the flesh leads to a spiritual existence. In the East, the approach is a much more holistic one. There are no differences or definitions of the good or bad sides of humanity. Zen advocates the comprehensive understanding of the whole. There is a synthesis. At a special point, the follower reaches the union it looks for, but not with God. The union is with mother nature.

To the outsider, Zen appears to be some kind of occult practice. The results, however, are there. And there is no magic involved, it's only that the methods are not as defined and clear as the Western mind would want them to be. Nothing is obvious. The methods and principles are totally intuitive; they are vague and ill-defined to say the least (far from clear, until the student grasps how the change happens in the form of an epiphany).

In actual fact, Zen is a highly refined kind of practical mysticism. It is the essence of Buddhism—what became of Buddhism after centuries of searching, until it reached Japan.

JAPAN AND ZEN

"How hard, then, and yet how easy it is to understand the truth of Zen! Hard because to understand it is not to understand it; easy because not to understand it is to understand it. A Master declares that even Buddha Sakyamuni and Bodhisatva Maitreya do not understand it, where simple-minded knaves do understand it."
D.T. Suzuki - *An Introduction to Zen Buddhism*

我と来て遊べや親のない雀
― 小林一茶
"Wareto kite asobe ya oya no nai suzume"
"Come play with me, little orphan sparrow."

Kobayashi Issa

遠山や目玉に写るとんぼかな
― 小林一茶
Tōyama ya medama ni utsuru tonbo ka na.
"Distant mountains
Reflected in the eyes
Of a dragonfly"
― Kobayashi Issa

"Just as the farmer irrigates a field, a fletcher fashions an arrow, and a carpenter shapes a piece of wood, so the sage tames his self."
Gautama Buddha - Dhammapada

Mahayana Buddhism made its first entrance into Japan in the sixth-century, but it was not a grand entrance; it was a strange one, and one that says tons about the nature of the doctrine.

The rulers of the Yamato Court, in Nara, had taken over all agricultural land in Japan and had managed to suppress the main clans. Their government was strong.

In 538 CE, the King of Baekje, one the small kingdoms of Korea, needed assistance from the powerful Yamato Court. With his delegation, the Korean King sent a golden image of Buddha, some parasols, and some sacred religious scriptures. The presents came accompanied by a letter in which the King expressed his admiration for the Buddhist doctrine: *"Among all the doctrines, this is the most excellent. It is hard to explain, though, and hard to understand"*. The Japanese reply was along similar lines: *"We have never had the opportunity to listen to such a wonderful doctrine. We do not seem to understand it ourselves either"*.

The emperor gave the image to the head of the Soga clan, so that he could enjoy it and worship in private. The Soga advocated the spread of Buddhism.

But, with a beginning such as it had, it is easy to understand why Buddhism did not spread straight away. There were also powerful nationalistic forces: two rival factions, the Mononobe and the Nakatomi clans opposed Buddhism and claimed the image had caused a plague. The temple in which the image was housed ended up being burnt and the image of Buddha, thrown into a canal by an enraged mob.

By 857 CE, however, the Soga had their revenge. They annihilated their rivals in battle, built the Hōkōji Temple (The Temple of the Rising Truth), and encouraged the spread of the Buddhist doctrine.

Somehow, Buddhism and Shinto (ancestor worship) learned to coexist. They both remain extremely popular and important to the country.

By the eighth century CE, a monk named Saichō founded Tendai Buddhism. Tendai is a school of Buddhism that incorporates Indian and Chinese traditions. It accepts the unity of all Buddhist schools. Tendai stresses the interconnectedness of all things and the Buddha nature of all beings. Tendai is the school that directly preceded Zen.

Saichō established the headquarters of Tendai in Mount Hiei (Hieizan), near Kyoto. It included three thousand buildings and twenty thousand monks. Unfortunately, the centre was destroyed in 1571. Saichō advocated education and co-operation with the state. As a result, Hieizan became the main Japanese centre of learning.

By the the thirteenth century CE, a monk called Eihei Dōgen, after studying at the monasteries of Hieizan, decides to abandon the teachings of Tendai Buddhism and becomes interested in Zen. The school (that in China was known as *Ch'an*) had been introduced into Japan by another monk, Eisai, who had learnt it in China. Following his steps, Dōgen goes to China, looking for enlightenment. After four years, he becomes enlightened and returns to Japan in 1227.

Throughout his life, Dōgen delivered several sermons that were compiled after his death under the title of *Kana Shōbō-*

genzō (True Dharma Eye). It was written in Japanese, as opposed to other works on *Ch'an*, which were written in Chinese. The legend is that the *Shōbōgenzō* is a distillation of the teachings transmitted by Buddhist Masters going all the way back to Gautama Buddha. The *Shōbōgenzō* refers to the way of awakening, i.e., the Zen way, that is not included in Pali texts. There are several versions of it.

According to some reports, Dōgen's first doubts had to do with training. If we all share the same Buddha Nature, why do we need to practice? He eventually realised what other animal species understand instinctively, you learn to build a nest through practice; you do not ask your parents how to hunt either. You watch, imitate, emulate. Then you become the hunter.

In Japan, Dōgen's methods are extremely successful. Everyone is free to practice Zen. The type of meditation he relied on in order to reach *satori* was *zazen*, which involves sitting cross-legged in silence. He added everyday chores to the practice of Zen. Keeping house, then, becomes a quasi-religious exercise.

With Dōgen's methods, Zen meditation becomes so prevalent in Japan, that it keeps on flourishing today.

In reality—and in line with the total simplification advocated by *Ch'an*—the main method Dōgen chose was to do away with teaching altogether. That was the most important refinement to Gautama Buddha's doctrine. No teaching, no explanation, no words. People who went to his school were forced to find their own ways from the resources of their own character. There was no prayer or ritual to aid them in their quest. All they needed was a zealous spirituality, faith that there was something beyond the self. Then, they had to focus on their meditation to achieve concentration and mindfulness; *satori* would emerge from all that.

True to the principles of Zen, Dōgen rejected all kinds of honours, including a purple robe, sent by the Emperor, which he never wore. According to the monk, who by then had founded the Sōtō sect, he didn't need official approval or encouragement; his school of Buddhism was far older than the Kamakura government—he argued—as it had been transmitted by generations of monks, one by one, beginning with Buddha himself.

Sōtō is one of two main Zen schools in Japan. The other one is Rinzai. There are small differences in the way their teachings are conducted, but both essentially maintain the traditions of Zen.

Eventually, Zen Buddhism came to rely on official favour. Emperors sought the advice of monks on many subjects.

Zen monks had a huge impact in Japanese life, not in terms of appealing to the masses, but in changing the habits of the Japanese higher classes. Many aspects of daily life were influenced by the monks: garden design, the tea ceremony, calligraphy and all types of art and literature became areas where the advice of Zen monks was almost essential.

Musō Soseki, for instance, a thirteenth century Zen master, born within the noble Masamura family, was a renowned poet and calligraphist, but he also became a referential figure in terms of garden design. He had studied the *Shingon* and *Tendai* schools of Buddhism but, after having a dream in which the founder of the *Ch'an* sect in China visited him, he decided to convert to Zen. His gardens include the usual trees, shrubs and greenery, but he also added special features like rocks and raked sand to symbolise the essence of human life, for instance. The intention behind those gardens was for them to encourage meditation. Rock gardens are famous in the West as an easily recognisable element of Zen Buddhism.

Zen Masters in Japan coincided with *Shingon* teachers that *satori* was the most important goal in life; however, following the example of Eisai and Dōgen, they insisted on the highly private and personal nature of enlightenment. Zen acquired its own views, separate from other Buddhist doctrines. The essence of Zen became something to be demonstrated rather than 'taught' in monasteries, in an atmosphere of quiet meditation. Among the main principles that reigned in those monasteries: 1) everyday chores are an important task of spiritual significance, and 2) thinking is useless.

Novices eventually find for themselves that reasoning becomes secondary and intuitive insight takes over. Nobody can reach enlightenment through logic and rationality. Zen cannot be actually taught by means of words—as may happen with other doctrines—it has to be demonstrated, but independent learning is the most important part of the process. Novices have to comprehend Zen by themselves. Nobody can do the learning for them.

So, daily occurrences are important, as is learning without help. Simplicity is one of the main principles of Zen, and that applies to Zen aesthetics as well. Nothing can be baroque, elaborate or artificial. Surrounding oneself with simple things is conducive to feeling, that is part of the philosophy behind rock gardens, or of the blank spaces that are typical of *Suiboku-ga* monochrome ink paintings.

Zen poetry, mainly a type of poem called *haiku*, reflects that simplicity:

> *The bee emerging from deep within the peony*
> *departs reluctantly*
> -Bashō

> *Like the little stream making its way through the mossy crevices I,*
> *too, quietly turn clear and transparent*

-Ryokan

Most Zen monasteries have a big room called the "Meditation Hall", or *zendo*. Monks spend a lot of their time there, as there is where they practice *zazen*, or cross-legged meditation. Each monk has his own space, a *tatami* mat, where he can meditate, and also eat and sleep. Tatami mats are approximately two meters in length by one meter in width. There, monks keep their belongings, which are few.

Each monk has a quilt, but no pillow. When he sleeps, he usually rests his head on his personal effects (robes, a few books, a razor, and some bowls, which he carries in a box). Having or wanting more possessions is considered bad for the spirit.

Meditation is not the main activity in Zen monasteries. Monks have to work. No chore is too menial for them, on the contrary, the lower the task, the better it is for their spiritual life. They have to sweep, scrub, cook, gather timber for fires, till the soil, or go begging in the villages that surround the monastery. No work means no eating. That is part of their philosophy of life. The concept of the working monk was introduced by Hyakujo, a Chinese monk in the eighth century BCE. Hyakujo's idea goes well with the spirit of Zen. Mental work, or any kind of abstract work, has no value for a monk.

Monks have two (sometimes three) meals a day. The meals are frugal, and consist of rice, sometimes mixed with barley, soup and pickled vegetables. Zen, however, does not value asceticism as such. It's just that the monk is supposed to use what he has and not waste anything. After the meal, there should be no grains of rice left in the bowl.

Monks are supposed to lead a simple life, in which humility and poverty are central. That doesn't mean, however, that they cannot enjoy fun or laughter. Their lives are industrious and ordered, and there is no room for useless suffering. The important difference with Western thought is that cognition means suffering.

MATERIALISM VS DUALISM

ROBERT LAWRENCE KUHN

I have discussed materialism many times in this book. I have discussed how it limits, how it constrains science in its study of human consciousness. Until science recognises that only one component of the human mind is biological, it will not be able to progress towards a plausible theory of consciousness.

In the *Introduction* I mention TV interviews as a source of information for the book. In one of them (of Robert Lawrence Kuhn conducted by Hilary Lawson), Kuhn—a public intellectual, arguably the person who knows the most about the different theories of consciousness—confirms that he is convinced (with a 92% probability of being right, according to him) that none of the current materialist theories is correct. Lawson asks the following question:

[HL: "So, let me begin, Robert, by asking you if you don't think that one, any one, of these theories, is more productive than another, ... er, and I do... is that right? Do you think they're all somehow viable?"]

He replies:

[RLK:] *"You know, I really have to think hard about that question. <u>I guess I can shortcircuit it by saying: it's not that we have too many theories, it's that we have one too few.</u> Er... uh... you know... I can put it that way, but every theory that I put in there, I felt something nice about. <u>I don't think any of them are right. I mean I don't think anybody has an answer</u>,* er... but everything has an interesting inside that gives me a different way.... and I've looked at this for 50 years or more now, and I looked at from every angle, and I've interviewed people on 'Closer to the Truth'..."*[1]

Lawson asks again:

[HL: *"Why don't you just give up on the idea that there is a right one? and er... But you're really attracted with the idea that there is a right one, it's just that we haven't got it yet?"*]

And Kuhn replies:

"And we may never be able to get it but I am, but I am... completely convinced that there is a right one. Whether it's possible to get it or not, that I'm not sure, I kind of doubt it. There are people who say...—and very distinguished people, friends and names that you know, who've appeared here—who would say that it's not possible for the human brain inside this universe to be able to understand consciousness. That's because we didn't evolve ... we evolved to escape the tigers on the savannahs of Africa... It's kind of a miracle that we can understand quantum physics and the theory of closure, ... how that happens... who knows? but we'll never be able to understand consciousness. It's just because of the limitations of our brains.

I reject that argument. I don't think that our brains are limited to where we couldn't understand consciousness. <u>I think the limitation of understanding consciousness has to do with this physical/men-</u>

---

1. Kuhn, R L – *A landscape of consciousness* – Interview conducted by Lawson, H – Retrieved from TV interview on YouTube, 8 August 2025. * My underlining.

tal... um ... um ... divide and... and is that achievable so that's why I would come down on a theory of consciousness that goes beyond the physical, so I would not be one to ...um to think that a purely physical theory would... er... would be true* *and I would, I would've given you that 98% certainty at some point in the past but one incident in my life… it would take a minute to explain was very personal change that percentage from 98% sure that materialism is wrong to maybe 92%..."*[2]

From what Kuhn is saying, it is obvious that most of the current theories of consciousness that have been submitted for approval by peers are based on purely materialist (physicalist) grounds. One of the aims of this book is to suggest to researchers that another possibility would be to reconsider Descartes and to reconsider substance dualism.

None of the current theories appear to take into account how language resulted in meaning—which did not exist before language— or how feedback from meaning resulted in cognition.

---

2. Kuhn, R L – *A landscape of consciousness* – Interview conducted by Lawson, H – Retrieved from TV interview on YouTube, 8 August 2025. * My underlining.

CONCLUSION

"I seem to feel Napoleon's influence on our quiet evening in the garden for instance — I think I see for a moment how our minds are all threaded together — how any live mind today is of the very same stuff as Plato's & Euripides. It is only a continuation & development of the same thing. It is this common mind that binds the whole world together; & all the world is mind."
Journal, 1 July 1903-
Virginia Woolf

Originally, proto-humans were born, like all other animals, biologically equipped with sentience but, like all other species, with only a germ of cognition; there was no division—I would guess—between the internal perception of the individual's own body and the perception of the rest of nature. An iteration of the species, the individual was one with all of reality. I suppose this is the way Eastern thought views 'the original mind'. There was hurt, there was pain, and affection, and instinct. Danger was perceived. Good food and sexual encounters were enjoyed but, in general, there was no analysis of reality. No divisions between 'I' and 'otherness'. No witnessing.

After hundreds of thousands of years as one animal species among many, our hominin ancestors developed—created and developed would be more appropriate—complex language. Renowned linguist Daniel Everett—the one who produced the Pirahã grammar—talks about the 'invention' of language. In any case, it took a long time.

Complex thought developed together with recursive language through a process that involved mutual feedback. It was a revolution. The way I see it, it was more than that: it was a meta-evolutionary event of unique proportions, only comparable to the origin of life on Earth. Humanity was established by adding that artificial layer of consciousness. Pre- and post-language layers co-exist to this day in our consciousness—that fact, I believe—is quite evident. Once you check the history, understanding consciousness becomes much easier.

The stage humanity has reached now is that of a superorganism that has actually taken over the planet. Eight billion individuals with a written history, buildings, transport, creations, beliefs, cultures, social institutions, etc. There are many obstacles, but we are so many now that globalisation and cultural integration, are eventually inevitable.

We are eight billion individuals who can communicate with each other, sometimes through interpreters or technology; but individuals who travel around the planet and to the moon, and are planning to travel to other planets. We represent different cultures, countries and languages. Humanity keeps on exploring. Humanity keeps advancing on its quest for knowledge.

Cognition seeks explanations; cognition wants to find out about its own environment (human consciousness). Thought, when expressed, analysed and verified, becomes knowledge. Knowledge is the aim of philosophy and science.

In human consciousness, sentience, the original layer, is the natural one. Ineffable, it requires no explanation for anything, basically because explanations are cognitive—a human artifice—and sentience is purely biological. We share it with our fellow animals, who need no explanations for their lives, behaviours or perceptions.

Western religion, philosophy and science provide different answers to some old questions and some new ones. Some answers may be true, some may be fiction, but they all are—and will be—limited to what is explainable.

Having failed to understand Wittgenstein and Schrödinger, but especially Wittgenstein, scientists and philosophers obstinately try to explain experience—what is worse, they fail to understand that human consciousness includes two discrete components: sentience and cognition. Paradoxically, trying to explain or measure sentience does not make any 'sense'.

The current scientific worldview appears to ignore that there are clear boundaries within consciousness; in that worldview, sentience and cognition are jumbled up: but the component that cannot be understood is only sentience. Cognition can be understood because it is the way we communicate. For humanity to be able to operate as humanity, individuals need to understand each other. Again, no other animal species requires understanding to that extent. Once that is seen clearly, consciousness is within reach.

The issues and the solutions posed by the brilliant minds I mention and cite in this book—and the conjectures I submit—are transcendental. Some of the conjectures may be highly

CONCLUSION

controversial and subject to criticism, maybe even ridicule. There are many more unanswered questions. We have dealt with some of them.

Children cannot learn to communicate without input from the parents and/or the collective. Toddlers need to be taught. They cannot learn by themselves.

If we accept that, then we accept that language is not something natural. It does not come with us as part of our biological equipment. Evidence of it is that we have to learn it every generation, and that there is no universal language.

We never question the fact because human beings have been partly artificial from the beginning of humanity. That happened tens of thousands of years ago. *H. sapiens [the man who knows],* is exactly that, the first hominin who can think, and we can think mainly because of language.

When our ancestors 'invented' language—as Everett would say—meaning came with it. We have mentioned before that the process that took us from grunts and yells to meaning and words and syntax took many thousands of years. In any case, the origin of our thinking skills is foggy but unquestionably true. One day we grunted, much later, we were speaking and we had become human. Given that it was such a long period, trying to determine the exact moment when an evolutionary process became meta-evolutionary and we could communicate complex thoughts would be an almost impossible task. We are sure, however, that one day our species transitioned from animals with a germ of thought to human beings with complex cognition.

There are other species that are partly artificial as well—like cats and dogs, poultry and cattle. They are not totally natural because we have bred secondary species derived from their original natural species. Cattle were originally wild auruchs; dogs were wolves, and so on, until we decided to alter their genetic material to suit our purposes. They are partly artificial because they are dependent upon our species; and we use them for transportation, food, etcetera.

In the case of our species, our distant ancestors pulled themselves up by their [non-existent] bootstraps and, from the animals they were, they became fully human. Going totally against nature (against normal evolution), they accomplished something that even today appears impossible.

A recent study has discovered a protein that—according to those conducting the study—may have sparked the origin of human language. That conclusion purports to explain that the ancestors of *H. sapiens* acquired language while *Neanderthals* and *Denisovans* did not because our species had a special, species-specific variant of the NOVA1 gene, called I197V. Why would we have that gene and the other *Homo* species not? The answer is that, although NOVA1 may have contributed to the evolution of human language, we did not have NOVA1 before 'inventing' language, if you like. It's the other way around: the gene came into being as a result of our ancestors' behaviour. Cells turn on or suppress certain genes as a result of their environment, in this case, the social behaviour of our ancestors. Gene activation often follows environmental changes. I would argue that language—and cognition, in a feedback loop—were part of a meta-evolutionary phenomenon that resulted in the origin of our species.

We have been preoccupied by the mysterious origin and nature of our consciousness for quite some time. How come we are the way we are? How are we different from other animals? Well, apart from the fact that we are alive like them,

CONCLUSION

many things make us different from them, if we want truisms: we wear clothes, we can communicate with each other, we can keep time, we are creative, we are artistic and adventurous, our cultures decide what is right and what is wrong, we have identities and free will, we understand abstract concepts, etc. We have divided ourselves into ethnic groups and geographic areas with which we identify, and which we call countries, we have governments of different types, religions, we trade, etc.

Other animal species are creative and adventurous, one might say. Yes, but the big difference lies in the nature of their creativeness, and their adventurousness. Their skills are totally instinctive. How can you tell the difference? Well, beavers build dams, but that is all they can build. They cannot build anything else. Of course, they wouldn't be able to build a tennis court. That would be too complicated. But you cannot ask a beaver to build a bridge with the same logs he uses for the dam either. He builds dams instinctively. The same thing happens with a bird's nest or a spider web. A swallow and other migratory birds travel long distances; a whale travels long distances. But they are not adventurous. They always follow the same routes for specific purposes. That is because their travelling is instinctive. Humans can choose where they want to go, or decide on different routes taking dangers or obstacles into consideration.

We have seen that animals have some degree of rationality, some degree of thought. Crows solve problems and use tools to obtain food. That means a germ of cognition. So do chimpanzees and other animal species. What they do not have is complex thought, recursive language or metacognition. They cannot speculate about their own thoughts, neither can they convey complex thinking to other individuals because they lack language.

I would say that neither Hume nor Descartes were right about the reasoning of animals. Descartes compared them to automata: no cognition at all. Hume said they were *"endowed with reason and thought"* like human beings: full cognition. The way I see it, the truth lies somewhere in between those two extremes. Animals cannot carry out complex tasks, neither can they communicate complex ideas because they cannot think conceptually, or abstractly, if you prefer. They lack language. As a linguist I would tend to say that human beings can think complex thoughts, solve abstract problems and co-operate on large enterprises because we have recursive language. Language is a unique phenomenon that has given us cognition beyond the limited degree other animal species have. What animals have is qualitatively and quantitatively negligible compared to human communication and cognition. Perhaps it should not be treated as full reasoning. I would argue that causal reasoning (the ability to make a tool in order to obtain something), based on the input of the senses, cannot be compared to how human beings understand the relations that may exist between thoughts. That really is cognition. Purely human cognition.

I submit here that it is highly improbable that AI will ever reach the level of human consciousness. AI cannot be sentient by definition. Its nature, as the name indicates, is totally artificial. Sentience is biological. Furthermore, large language models like ChatGPT, do not actually 'understand' language. ChatGPT is a logarithm with a massive memory. Operators input millions of sentences and the logarithm statistically learns the exact words to use within a certain context. It's a Bayesian process. That does not mean it thinks or understands. The logarithm can also fake being sentient, but that comes a result of input. It cannot begin to do anything without outside input because it has no agency.

In order for an automaton to achieve something similar to human consciousness, it would need to be hybrid, like us. Human consciousness requires cognition, but it also requires the live biological component.

In this book I attempt to prove that the components of our consciousness—its layers—have different natures. Why do I say that? When we analyse why, we find that sentience—originally the sole component of our consciousness, like those of other animal species—is natural. Cognition, however, is the artificial creation of our human ancestors, the result of a 'black swan' process of meta-evolution. From our perspective, human consciousness appears to be a hybrid phenomenon.

The irony of it all is that we have always been what we have long feared we would become: artificial, or at least partly artificial. Human beings are not natural animals. We have an extremely important artificial component, which is our cognition. And the only way we could have developed cognition is through language.

Perhaps we could say that this book amounts to a basic linguistic theory of consciousness. Truth is, language is the gel that binds humanity together. We cannot live without it or be human without it. It so happens that, without language, human consciousness would not exist.

ACKNOWLEDGMENTS

My wife Inés, as usual. She read all versions, corrected, marked typos, suggested changes. The job of an editor, but through all the painful stages of the book.

John Watts, former diplomat, a man with a solid educational background and a good deal of experience of the world, but more than anything a good friend, read the book, suggested changes, and was kind enough to write the Foreword.

My brother, Patocho, read the final version and made several suggestions, especially about addressing it to a wider audience and specifying that, even if that wasn't the original intention, the book amounts to a linguistic theory of consciousness.

Rodrigo Suárez, biologist, Associate Professor, University of Queensland, patiently listened to the ideas when the book was still meant to be a thesis, discussed, and provided a great deal of help.

Madeleine Beekman, biologist, Professor Emerita, University of Sydney, listened to ideas and proposed some changes. The dialogue with Professor Beekman is still ongoing.

I owe all of them a huge debt of gratitude.

www.ingramcontent.com/pod-product-compliance
Lightning Source LLC
Chambersburg PA
CBHW071239070526
44583CB00017B/2254